Fundamentals
of Protection and Safety
for the Private Protection Officer

Robert J. Meadows

California Lutheran University

Prentice Hall
Englewood Cliffs, New Jersey 07632

Library of Congress Cataloging-in-Publication Data

Meadows, Robert J.
  Fundamentals of protection and safety for the private protection
officer / Robert J. Meadows
     p.   cm.
  Includes bibliographical references and index.
  ISBN 0-13-720509-0
  1. Police, Private—United States.   2. Private security services—United States
I. Title.
HV8291.U6M43   1995
363.2'89'0973—dc20   94-22259
                                                             CIP

Editorial/production supervision, interior design,
   and electronic composition: *Julie Boddorf*
Cover design: *Todd D. Ware*
Electronic art production: *Julie Boddorf/TTP International*
Director of production and manufacturing: *Bruce Johnson*
Managing editor: *Mary Carnis*
Production coordinator: *Ed O'Dougherty*
Acquisitions editor: *Robin Baliszewski*
Editorial assistant: *Rose Mary Florio*

© 1995 by Prentice-Hall, Inc.
A Simon & Schuster Company
Englewood Cliffs, New Jersey 07632

Printed in the United States of America

10   9   8   7   6   5   4   3   2   1

ISBN   0-13-720509-0

PRENTICE-HALL INTERNATIONAL (UK) LIMITED, *London*
PRENTICE-HALL OF AUSTRALIA PTY. LIMITED, *Sydney*
PRENTICE-HALL CANADA INC., *Toronto*
PRENTICE-HALL HISPANOAMERICANA, S.A., *Mexico*
PRENTICE-HALL OF INDIA PRIVATE LIMITED, *New Delhi*
PRENTICE-HALL OF JAPAN, INC., *Tokyo*
SIMON & SCHUSTER ASIA PTE. LTD., *Singapore*
EDITORA PRENTICE-HALL DO BRASIL, LTDA., *Rio de Janeiro*

*This book is dedicated to my wife Glenna
and sons James, Conrad, and Garrett.*

Contents

Chapter

3 PATROL PROCEDURES   34

Appendix

A CODE OF ETHICS FOR PRIVATE SECURITY EMPLOYEES

Appendix

B HOURLY WAGES FOR PROTECTION OFFICERS
FOR SELECTED CITIES

Appendix

C 180-HOUR FIVE-WEEK SECURITY OFFICER COURSE

Appendix

D STATE STATUTES REGULATING SECURITY GUARDS

INDEX

Preface

The private protection business is expanding at an incredible pace. More persons are employed in private protection services than in federal, state, and local law enforcement agencies.

This book was written for the operational-level private protection officer. Protection officers are employed in a number of enterprises. Their responsibilities range from fixed-post assignments, to roving patrols in commercial and residential settings. Protection officers provide services to manufacturing plants, hospitals, schools, hotels, entertainment establishments, residential complexes, and many other businesses.

This book is intended to meet the general training needs of the protection officer regardless of assignment. The chapters are divided into relevant topical areas and provide practical information for the working protection officer and Supervisor. Protection practices may vary depending upon business needs; however, there are some fundamental training requirements that every protection officer should fulfill.

As society becomes more compartmentalized, as long as crime remains, and as long the public fund for public protection continues to dwindle, the need for private protection will become more crucial.

This book recognizes that private protection involves more than simple observation and reporting skills. The officer must be familiar with legal principles, language skills, report writing, health and safety, first aid response, and other relevant topics. Private protection training is seriously lacking in both availability and substance.

The best use of this book would be in a classroom situation as part of an instructional program in private protection. Users of the book will benefit through interaction with other students. Each chapter has learning exercises designed for classroom discussion. It is recognized that some topics can be expanded, while others require demonstration. It is recommended that additional instructional materials be introduced to augment the book topics.

The success of the book depends upon response from the users. I encourage suggestions for future revisions.

ACKNOWLEDGMENTS

Writing a book requires the advice and assistance of others. The preparation of this book represents countless hours of research, writing and rewriting, and consultation with protection professionals.

Among the practitioners and academicians who helped in preparation of this book, the following deserve special recognition: Bob Breardsley, of the Public Safety Training Association, San Diego, California; John Chuvala, CPP, Western Illinois University, Macomb, Illinois; Jack Egger, Director, Studio Protection and Disaster Preparedness, Warner Bros. Studios, Burbank, California; Lt. A.J. "Art" Farrar, Ventura Public Department, Ventura, California; Robert G. Lee, CPP, CDRP, Vice President Emergency Planning/ Corporate Security Environment Affairs/ Safety, Great Western Financial Corporation, Chatsworth, California; James Loriega, Municipal Training Center, Brooklyn, New York; Alan T. Mather, Captain, military Intelligence Officer, Fayetteville, North Carolina; John J. O'Kane, Adirondack Community College, Queensburg, New York; Gary Ward, Director of Security /Parking, Children's Hospital of Orange County, Orange, California.

In addition, I would like to extend a special thanks to the numerous contract and proprietary protection officers I consulted with throughout the country. Their practical advice and assistance was invaluable.

An extra special thanks is given to Tajiana Standish and Kathy Chen for their computer expertise and technical advice in preparing the book. Their many hours of work is greatly appreciated.

Finally, I would like to thank Robin Baliszewksi of Prentice Hall Education, Career & Technology for her caring and professional guidance in getting this book into publication. Thank you all.

ABOUT THE AUTHOR

Robert J. Meadows is an associate professor in the Department of Sociology and Criminal Justice at California Lutheran University, Thousand Oaks, California. Prior to teaching, Dr. Meadows was a security investigator and manager for May Company department stores and a police officer for the city of Los Angeles. Dr. Meadows is a recognized consultant and expert witness on issues pertaining to premises liability for negligent security. Dr. Meadows received his bachelor's degree from Northern Arizona University, master's degree and doctorate of education degree from Pepperdine University, and Ph.D in Criminal Justice from Claremont Graduate School. Dr. Meadows is a certified protector professional (CPP) and a certified Fraud Examiner (CFE). Dr. Meadows has published numerous articles on policing and private security.

Chapter

1

Overview of the Private Protection Function

Learning Objectives

After studying this chapter, you should be able to explain the following:

Categories of private protection
   services
Code of ethics for the private
   protection officer
Contract security
Information security
Personnel security
Physical security
Private investigations and private
   investigators

Proprietary security
Reasons for the growth of private
   protection
Risk management
Role of private protection
Security and safety
Types of private protection industries

Private protection is becoming a major growth industry.[1] The public demands the police focus on crime prevention and offender apprehension. However, the police cannot provide the necessary protection for everyone; there are serious protection needs requiring the attention of those who are trained in security and safety measures. Private protection therefore involves security of areas not normally patrolled by the police, and offers special services and equipment not provided by the police. The private protection industry outnumbers public law enforcement by a ratio of nearly 3:1. In other words, there are nearly three working protection officers for every law enforcement officer. As pointed out by the National Institute of Justice, "Private security is now clearly the nation's primary protective resource, outspending public law enforcement by 73 percent and employing 2 1/2 times the workforce." The business of private protection is comprised of the following categories of services:

- *Proprietary (in-house) security.*
  Protection officers hired and controlled by the protected organization.
- *Contract guard and patrol services.*
  The hiring of a guard or patrol service to protect an organization for a fee.
- *Alarm services.*
- *Private investigators.*
  Investigator gathers facts and background information for use by insurance companies, attorneys, various businesses, and private parties
- *Armored car services.*
- *Manufacturers of security equipment.*
  For example, locking devices and alarms.
- *Locksmiths.*
  Persons trained in the use of locking devices.
- *Security consultants and engineers.*
  Professionally trained security experts.
- *Other services.*
  Bodyguards, escort services, canine patrols

The growth of private protection can be attributed to four major causes:

1. *Rising work-place crime.* Computer fraud and assorted technical crimes are becoming a problem.
2. *Fear of lawsuits for inadequate protection.* Suits filed against businesses for third-party crimes are increasing, forcing businesses to consider increased security needs.
3. *Less government spending on public protection.* Criminal justice budgets are decreasing due to declining tax bases.
4. *Growing realization of protection measures.* There are many protection products on the market providing more public awareness of security options for personal and business protection.

Another function of private protection is undercover or private investigation. In many businesses, protection officers have the responsibility of seeking out dishonest employees and shoplifters. In businesses where employee drug

---

[1]For the purposes of this book, the term *protection officer* will be used to encompass the broad security and safety functions.

usage is a problem, undercover officers may be assigned the responsibility of identifying users and reporting them to management. These private agents may be hired by the concerned business, or an outside contractor.

Private investigators, or "private eyes" as they are sometimes referred to by the media, are persons who work independently or with a corporation. Some states regulate the business of private investigators. In California, for example, a state license is required in order for an individual or business entity to begin operating as an independent private investigative business. To be eligible to take the California state licensing exam, the prospective investigator must have accumulated several thousand hours of investigative experience and successfully completed a state background check. Private investigators perform a wide range of duties such as investigating insurance claims, employee background investigations, case investigations for attorneys, and surveillance operations.

MEASURES OF PROTECTION

The terms *security, safety, risk management* and *loss prevention* are used in discussing the role of private protection. All involve the process of protecting human and property resources. Protection is a proactive service. In other words, private protection provides a broad range of services to a clientele composed of individuals, institutions, businesses, and some governmental agencies. A number of authors have described these services in terms of three functional areas: (1) information security, (2) personnel security, and (3) physical security. Services performed in these functional areas include gathering information, maintaining order, and protecting persons and property through detection and prevention of crime, as opposed to reacting after a crime has occurred.

Information security includes those measures required to protect the confidentiality of information owned or held, such as

1. Grading information according to its sensitivity and affording protection accordingly.
2. Providing physical protection.
3. Educating employees as to individual responsibilities.
4. Screening of employees and visitors.
5. Controlling access and personnel identification.

Personnel security includes measures necessary to protect employees or invitees of a facility from the effects of

1. Hostile propaganda and subversion.
2. Disloyalty.
3. Industrial espionage.
4. Fires and other disasters.
5. Strikes, riots, and other disturbances.
6. Injury and harassment.

Physical security encompasses measures necessary to protect a facility against the effects of unauthorized access, theft, fire, sabotage, loss, or other intentional crime or damage. Some of these measures include

1. Prevention of unauthorized access by means of security officers, barriers, fences, lighting, and alarms.

2. Control of authorized entry by personnel identification.
3. Prevention of employee crime and pilferage.
4. Fire prevention and control.
5. Prevention of accidents.
6. Implementation of traffic control and parking regulations.
7. Implementation of security surveys.
8. Control of locks, keys, and safes.
9. Control of materials.
10. Procedures of control.
11. Emergency measures.

As indicated in Figure 1–1, the three functional areas of protection overlap in the protection process. In other words, all three must be considered in business protection operations.

A SHORT HISTORY OF PRIVATE PROTECTION

Throughout history, security operations have existed in one form or another. Security was traditionally the responsibility of family members, a clan or tribe, since there were no organized government forces to provide protection. The use of private armies and hired guardians provided the only security for wealthy Americans. The American colonists borrowed from the English system of law enforcement; many existing American laws and law enforcement principles are taken from the English legal system. However, as the American population grew and became more diverse, new approaches to public safety were needed. The origins of contract security in the United States have usually been credited to Allen Pinkerton. As the railroad lines moved into sparsely populated areas, trains were attacked by outlaws. In 1855, Pinkerton began to provide protection for a number of railroads. The Pinkerton agency expanded and gradually provided security services to business, industry, and the United States Army during the Civil War. In years following, other private security agencies entered the field of private protection. The William Burns International Detective Agency came into being in 1909, and the Wackenhut Corporation started doing business in 1954.

Other businesses began to form their own in-house security services rather than contracting with other agencies. Today, private protection is offered through a number of contract guard services and corporate (in-house) personnel. Private protection has emerged as a major force in safeguarding others from crime. The companies listed in Table 1–1 represent the major protection

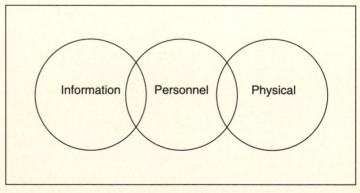

FIGURE 1–1

TABLE 1–1

Company/Head Office	Employees	1988 Revenues
Pinkerton's, Inc ., Van Nuys, California	55,000	$ 652,000,000
Burns International Security Services, Paramus, New Jersey (estimated)	30,000	$ 435,000,000
The Wackenhut Corporation, Coral Gables, Florida	35,000	$ 400,000,000
Wells Fargo Guard Services, Parsippany, New Jersey	21,500	$ 250,000,000
American Protective Services, Oakland, California	9,000	$ 151,000,000
Globe Security, Deerfield Beach, Florida	10,000	$ 125,000,000
Stanley Smith Security, Inc., San Antonio, Texas	8,900	$ 120,000,000
Guardsmark, Inc., Memphis, Tennessee	8,000	$ 120,000,000
Allied Security, Inc., Pittsburgh, Pennsylvania (estimated)	6,000	$ 76,000,000
Advance Security, Inc., Atlanta, Georgia	6,000	$ 75,000,000
Totals	189,400	$ 2,404,000,000

Notes: Prepared from questionnaire responses for the *Security Letter Source Book* for revenues through December 31, 1988, except where "estimated" is indicated. Totals may include investigative and consulting services and also some revenue reflect sales of non-guard-related revenues.

Source: Security Letter Source Book, 1990–1991.

businesses in the United States. The emergence of a number of professional security organizations underscores the movement to professionalism. The American Society of Industrial Security (ASIS) which has a membership of over 24,000 security professionals, the International Association of Hospital Security (IAHS), and the National Association of School Security Directors (NASSD) are just a few such organizations. Protection officers can join an organization called the International Foundation for Protection Officers (IFPO). This organization offers a correspondence course which awards officers the distinction of Certified Protection Officer (CPO).[2]

THE PROTECTION ROLE

The role of private protection is extensive. Many private settings are confronted with problems similar to those found in the city. Thieves, muggers, and drug offenders are just a few of the many threats. Whether the property is a shopping mall, parking lot, hospital, school, or industrial plant, the protection officer must understand the importance of protecting others from crime, safety hazards, and natural disasters. In other words, a protection officer is concerned with risk assessment and response. Since the protection officer is in frequent contact with the public, he or she must be courteous and helpful. Everything an officer does should work toward building good will for the officer and the officer's employer. The code of ethics for private security employees serves as a reminder of the sensitive duties and responsibilities of a protection officer (see Appendix A).

Asset protection requires the ability to observe and report incidents which may involve crime or safety violations. The assets the protection officer guards are people, property, and information. This protection includes employees, suppliers, patrons, visitors, buildings, equipment, machinery, supplies, files, records, and other confidential information. A protection officer may be highly visible unless required to work undercover. A highly visible uniformed protec-

---

[2]Inquiries regarding this program should be addressed to: IFPO, Mount Royal College, Faculty of Continuing Education and Extension, Alberta, Canada. There is also a CPO program administered out of Bellingham, Washington.

tion officer can be a strong deterrent to criminal activity, since many intruders and criminals do not want to be detected. Protection officers may be expected to stop and question suspicious persons and in some cases initiate an arrest. The short snapshot roles described in this chapter offer evidence of the similarity of private protection roles regardless of the business.

The three aspects of the private protection role are identified in Figure 1–2. As indicated, the protection role may involve not only security, but also safety and emergency planning and response. It is recognized that some businesses require different levels of protection. In some manufacturing settings an officer's duties may be limited to access control, while in other businesses an officer may be expected to perform safety inspections as well as crime prevention duties. Some businesses require frequent contact with the public, while others may have minimal contact. Obviously, the officer's job description or post orders will determine what training is required. It is the responsibility of the protection officer's employer to assure that the protection officer is prepared to perform assigned duties and respond to a variety of incidents.

Unfortunately, the private protection business has had its share of negative publicity. The industry has employed guards who have victimized those they were hired to protect (Behar, 1992). The reasons for this is that some protection companies fail to conduct the proper background investigations in hiring protection officers. Another reason is poor training. However, it should be recognized that public law enforcement with its higher standards of screening and training of police officers also experiences problems in personnel misbehavior or mal-

A Snapshot: Security Duties in a Government Scientific Laboratory

The complex resembles a small city consisting of approximately one hundred buildings, including two off-lab satellite facilities and an annual operating budget approaching one billion dollars. Research and development for the National Aeronautics Space Agency, including Space Flight Operation Facility for the unmanned space flight programs are currently being done. Other projects for the state, local and other federal agencies are also done and they are too numerous to list. For example, the instrumentation for seismic studies of the San Andreas Fault are performed there for the State of California, including computer design systems for the Federal Aviation Agency (F.A.A.). This facility specializes in several areas, but is primarily known for its unmanned space programs such as "Voyager," "Galileo," and "Magellan."

Each day a security officer rotates in the duties assigned. These duties might include being posted at an entrance gate checking in/out all contractors/vendors and employees that enter daily, including randomly searching vehicles leaving the premises for computer hardware leaving the lab for other destinations. Desk jobs involve observing and reporting people going in and out of "closed" or "limited access" buildings. Mobile units check perimeter fencelines and other buildings off-lot after normal working hours. Each officer spends time with a senior training officer on learning the procedures and policies including a working knowledge of all area buildings; e.g., what's in them, who to contact for different types of circumstances confronted, etc.

Basic training consists of basic law that enables us to arrest someone, carry firearms, and enforce federal laws if violated by somebody in the lab. It's a general requirement to possess a current secret or top secret clearance from the Department of Defense. Because of the clearance requirement most security officers have had prior law enforcement, military police, or defense contractor security experience. Continuing education in administration of justice has been encouraged by management. Courses are being taught both on-lab and off-lab at the local Community College. Security is an in-house operation. Most security officers stay, while others leave to pursue careers in law enforcement or security management.

SECURITY
Physical Security
Crime Prevention
Executive Protection
Internal Investigations Including Undercover Operations

SAFETY
Accident Prevention
Hazardous Materials Recognition
OSHA Regulations
General Industry Safety Orders

EMERGENCY PLANNING
Readiness/Preparedness for Unusual Incidents
Emergency Response to Unusual Incidents

FIGURE 1–2   The protection role.

practice. The reputation of any business begins with the individual officer and the philosophy of the officer's employer.

Protection Officer Training

The business of private security is expected to grow and outspend law enforcement in the future. Law enforcement is limited in the amount of public protection it can provide. In other words, there are geographical and budgetary restraints affecting law enforcement services; therefore, private security is expected to grow as the public demands more protection (see Figures 1–3 and 1–4).

Protection officer training varies from state to state. Regardless of a state's requirements, employers may require more training than mandated by a particular state. Protection officers should receive continuous training; however, this is often the exception rather than the rule. There are no federal requirements for protection officer training, and few state requirements. However, some states require training in firearms, laws of arrest and other critical areas. The data in Appendix D provides a listing of state training standards for protection officers.

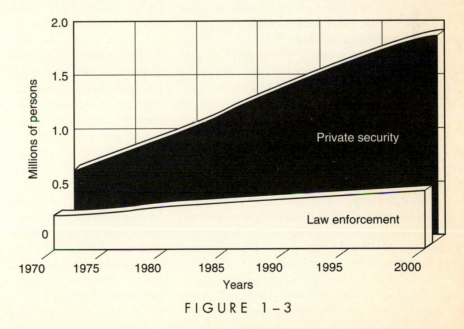

FIGURE 1–3

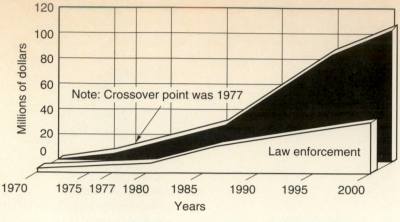

FIGURE 1-4

Note that in some states there are no statutory requirements for training, while in other states there are requirements. Some states specify in-house exclusion for training officers. However, since this table was produced, some states are seeking to upgrade training standards.

In addition to minimal state requirements, there are recommended standards for protection officer training. *The Private Security Task Force Report*, published in 1976, recommended the following:

Any person employed as an investigator or detective, guard or watchman, armored car personnel or armed courier, alarm system installer or servicer, or alarm respondent, including those presently employed part-time personnel, should successfully:

A Snapshot: Shopping Mall Protection

While observing and reporting is a standard to be practiced, active participation is expected as well. The mall consists of a movie theater, medical plaza, office complex, drug store, supermarket, six major banks, five bars, seven restaurants and a dance club.

The local city council has passed several city ordinances allowing protection officers to write city parking citations and to enforce reckless driving and exhibition of speed on mall property.

The mall has two patrol vehicles. Each protection officer carries a portable radio, and each unit has a radio, scanner, and air horn. Protection officers are armed with tear gas, batons, and firearms.

There are a number of duties protection officers are expected to perform. Some are locking and unlocking the mall, observing and reporting hazards in the mall such as liquid spills, or anything else that could be viewed or construed as a liability for the company.

Protection officers must also check and regulate the fire riser. Check valve pressure, and drain and fill the fire sprinkler system. Officers assist tenants with escorts to their cars and bank deposits.

Training consists of state-mandated Department of Consumer Affairs baton and firearms training and powers of arrest. This initial training totals approximately 24 hours. Other continual training provided by the company includes seminars such as verbal judo, police report writing, patrol procedures, gang seminars, drug recognition, various shoplifting and burglary colloquiums at local police departments.

Continual in-house training is provided, involving building searches, handcuffing techniques, and role playing. Each protection officer must qualify with his/her firearm quarterly. First aid, CPR, and fire control are also provided as training. Protection officers are required to know common Criminal Code sections and police radio codes.

1. Complete a minimum of 8 hours formal preassignment training;
2. Complete a basic training course of a minimum of 32 hours within 3 months of assignment. A maximum of 16 hours can be supervised on-the-job training.

The Public Training Officers Association of San Diego has developed an excellent five-week, 180-hour training course. A number of contract protection firms and corporations send their officers through this training program (see Appendix C). There are a number of community colleges and security businesses which offer training. It is recommended that protection officers seek out these opportunities. The Certified Protection Officer Program (CPO) discussed earlier in this chapter is another excellent opportunity to obtain protection training. This training does not require any classroom attendance.

The Gore and Martinez Training Proposals. Although a number of states mandate protection-officer training, the protection industry lacks regulation; however, there are federal legislative efforts to standardize protection-officer training.

One such effort, referred to as the Gore Bill, was named after its author Albert Gore, former Senator of Tennessee and Vice-President since 1993. The 1991 legislation is designed to standardize protection-officer training for those officers working in government security. The training would cover the following areas:

- Fire protection and prevention
- First aid
- Legal issues
- Investigation and detection
- Building safety
- Methods of handling crisis situations
- Crowd control
- Use of protection equipment
- Report writing

A Snapshot: Protection for a Government Defense Contractor

Protection officers are trained in a regimented manner. They receive certified Red Cross and CPR training. A Department of Defense (DOD) background check is requested. The background check can take up to six months. Next, the on-site training begins. Protection officers must view six hours of videos covering basic report writing, public relations, and legal rights and responsibilities. In addition, there is supervised (on-the-job) training in basic duties such as traffic control, lobby reception, and dock receiving.

Officers are taught patrol and lockup procedures for the building. The patrols are done frequently, but at random times and in several different patterns with a laser barcode reader. Lockup procedures include locking, arming, and testing the doors and vaults.

The officer's primary post is in the ground-level fire, life, safety, and security command center. When a door alarm, vault alarm, computer-system alarm, or motion detector is activated, officers are dispatched to investigate, and (if instructed) contact the police.

During a fire, earthquake, or other emergency, officers are authorized to contact the proper authorities and begin evacuation procedures without waiting for supervisor approval or the arrival of the fire department and/or police. The protection officer utilizes the command phone to simultaneously contact all fire phones.

The bill would mandate examination and certification procedures to ensure the quality of training.

Another legislative effort was initiated in 1992 by United States Representative Mathew Martinez (Democrat California). This piece of legislation is referred to as House Bill 5931. The Martinez Bill is more specific than the Gore Bill. This bill would mandate a minimum of eight hours of classroom instruction and successful completion of a written examination. The bill would further require four hours of on-the-job training for protection officers.

Individual states would set standards for protection-officer trainers. The Martinez Bill would include instruction in the following:

- Legal powers
- Safety and fire detection
- Report writing
- Patrol procedures
- Deportment and ethics
- Equipment use and other general information

Both bills are aimed at the unarmed protection officer. Additional measures are needed for armed officers. Only time will tell if these bills meet with success; however, it is crucial that protection officers, regardless of assignment, receive mandated classroom and on-the-job training.

COMPENSATION

The wages and benefits protection officers earn depend upon a number of factors. The amount of earnings may include the type of business protected, whether firearms are required, and the metropolitan area in which the protection officer is located (see Appendix B).

Generally, proprietary (in-house) protection officers earn more than contract officers. This is due largely to the fact that wage laws have been set by collective bargaining units. Contract officers usually receive less pay and benefits and also experience higher turnover rates. Unfortunately there are a number of protection firms which operate on shoestring budgets and provide minimal or no training. Salary rates may also vary according to the protection officer's experience and training, and the employer's budget and generosity.

SECURITY AND LOSS-PREVENTION OPPORTUNITIES

Employment opportunities in private protection are growing. As the need for private protection increases, there will be more demand for quality protection services. Law enforcement cannot provide the needed protection for business invitees. The following list is a sample of some industries using protection officers. It is recognized that some of these areas of employment require higher recruitment and training standards than others; however, they all require officers with initiative and responsibility.

Banking and Financial Institutions
General Banking Institutions
Bankcard Centers
Savings and Loan Companies
Financial Centers

Computer Security
Banks
Telephone Companies
Insurance Companies
Credit Card Companies
Crime Prevention

Credit Card Security
Banks
Retail Stores
Gasoline Companies

Educational Institutions
Universities
Junior/Community Colleges
Major School Districts

Fire Resources and Management
Small Communities
Major Firms, Aerospace

Health Care Institutions
Hospitals
Convalescent Homes
Retirement Communities
Pharmaceutical Firms

Nuclear Security
Production Facilities
Power Plants
Transportation Facilities

Public Utilities
Departments of Water and Power

Telephone Companies
Major Power Generating Facilities

Restaurant and Lodging Establishments
Large Restaurant Chains
Hotel and Motel Chains

Retail Stores
Department Stores
Grocery Stores
General Merchandise Stores

Transportation and Cargo
Airlines
Trucking Firms
Special Couriers
Railroads
Ports/Maritime Security

Executive Protection
VIPs
Diplomats
Celebrities

Conclusion

The role of protecting others is both challenging and rewarding. It requires persons who have patience, enjoy working with people, have the ability to communicate both orally and in writing, possess keen observation skills, and can work without close supervision. Protection officers must be alert, physically fit, and of high moral character. With the decline of public law enforcement services in some communities and the accompanying need for increased private protection, the business of private protection is bound to flourish in years to come. Private protection is a proactive service rather than a reactive force. In other words, private protection requires a prevention-oriented philosophy. Officers need to be alert and demonstrate a service orientation to others. The role of protection is also expanding to include risk-assessment issues such as health and safety inspections in the work place.

Discussion Questions

1. What is the difference between proprietary and contract security services?
2. Describe the importance of information, personnel, and physical security measures. How do these measures contribute to asset protection?
3. Discuss reasons for the growth of the private protection industry.
4. Identify and discuss the three aspects of the private protection role.
5. In what ways is the role of private protection similar to the role of public protection (policing)?
6. What can be done to improve the image of private protection officers?
7. How can the police and the private protection industry become partners in the business of crime prevention? What strategies could be used to develop such a partnership in your community?
8. There are a number of industries that employ private protection officers. In general, what should be the common hiring and training requirements for protection officers in all industries? Are there similar rule requirements regardless of industry?

9. Review the Code of Ethics in Appendix A. Can you think of any situations in which any of the ten ethical principles have been violated by your employer or fellow officers? Is it possible to be totally ethical in private protection? Explain.

References

BEHAR, RICHARD, "Thugs in Uniform," *Time*, March 9, 1992, pp. 44–46.

CUNNINGHAM, WILLIAM C., AND JOHN J. STRAUCHS, "Security Industry Trends: 1993 and Beyond," *Security Management*, February, 1992, pp. 16–23.

PAINE, D., *Basic Principles of Industrial Security*, Madison, WI: Oak Security Publications, 1972, p. 36.

POST, RICHARD S., AND ARTHUR A. KINGSBURY, *Security Administration: An Introduction.* Springfield, IL: Charles C. Thomas, 1970, p. 6.

"Private Security: Patterns and Trends," National Institute of Justice, U.S. Department of Justice, Washington, D.C., 1991.

"Private Security, Report of the Task Force on Private Security." National Advisory Committee on Criminal Justice Standards and Goals, U.S. Department of Justice, Law Enforcement Assistance Administration, Washington, D.C., 1976, pp. 5–6, 98.

URSIC, H. S., AND L. E. PAGANO, *Security Management Systems*, Springfield, IL: Charles C. Thomas, 1974, p. 90.

Chapter
2

Legal Aspects of Private Protection

Learning Objectives

After studying this chapter, you should be able to explain the following:

Arrest and detention
Assault
Battery
Bill of Rights
Burglary
Civil liability
Common law
Crime of omission
Crime of commission
Criminal liability
Criminal intent
Determination of negligence
Disorderly conduct
Exclusionary rule
Plain view doctrine
False arrest and imprisonment
Felony
Government of checks and balances
Indecent exposure
Intentional torts
Kidnapping
Larceny

Miranda rights
Misdemeanor
Murder
Negligent hiring
Negligent supervision
Negligent retention
Negligent training
Negligent torts
Probable cause
Public and Private Law
Public order crimes
Rape
Respondeat superior
Vicarious liability
Search
Seizure
Sources of law
Theft
Trespassing
Vandalism and malicious
    destruction of property

The protective role of security often invites involvement with the law; an understanding of legal limitations and expectations are of crucial importance if mistakes and liabilities are to be avoided. This chapter focuses on general security legal obligations regardless of assignment.

Protection officers and security supervisors should have a basic knowledge of law in order to minimize exposure to civil or criminal actions. This chapter will acquaint the protection officer with the sources of American law; the basic principles of criminal and civil law, arrest and detention, search and seizure; and the sources of civil liability. It is recognized that jurisdictions differ with regard to the definition of certain laws. The purpose of this chapter is to present a general understanding of law rather than to deal with any specific jurisdictions. Reference is made to the model penal code in explaining some common crimes. The code is used by a number of jurisdictions. It is recommended that protection officers consult with their supervisors, local police, or company attorneys regarding legal procedures operating in specific jurisdictions. Remember, laws sometimes change or are otherwise amended due to court decisions or legislative acts.

SOURCES OF LAW

The American legal system is based upon what is referred to as English *common law*. The common law refers to those customs, traditions and judicial decisions that guide courts in decision making. Common law is often referred to as *judge-made law* because initially rules were determined by judges, not by legislative or executive bodies. Much of the early common law found in England for crimes such as murder and larceny has been written into state statutes and penal codes.

In the United States, laws are enacted by representatives of the people. Laws are enforced, administered, and interpreted by civil servants and elected officials. Laws can be either substantive or procedural. Substantive law defines misbehavior or criminal acts, as well as the punishment for such offenses. State penal or criminal codes are a source of substantive law. Substantive law deals with crimes such as robbery, which is generally defined as "the unlawful taking of personal property through force or the fear of force." *Procedural law* refers to those rights and protections offered by the federal and state constitutions. An example of these protections is the *Bill of Rights*, the first ten amendments of the Constitution. Some of the rights guaranteed by the First Amendment are the freedoms of speech, religion, and the press. Under the Sixth Amendment, anyone accused of a crime has the right to an attorney regardless of his or her financial situation.

In our system of government, there are *three* branches of government.

1. *Legislative Branch.* Laws are enacted by the legislative branch. This branch includes state assemblies and the United States Congress. Various types of laws are enacted by these elected bodies. The laws of some states, for example, mandate specific training for protection officers.
2. *Executive Branch.* The executive branch administers and enforces laws. Governors can order national guard troops to quell riots and civil disturbances. The president can order federal troops to enforce federal court orders. Law enforcement agencies are charged with the responsibility of enforcing criminal laws.
3. *Judicial Branch.* Persons charged with crimes have the right to be tried before a judge or jury. Judges have the power of judicial review in deter-

mining the constitutionality of laws or ordinances. In other words, courts determine the fairness of the other two branches. Courts exist at the state and federal levels, with the highest court of the United States called the Supreme Court.

The three branches of government are referred to as a *government of checks and balances* because they are designed to assure that no one branch becomes too powerful. The legislature makes laws, the courts interpret laws, and the executive branch enforces laws. For example, the Fourth Amendment right against unreasonable searches and seizures often requires court interpretation as to the standard of reasonableness. As discussed later in the chapter, early U.S. Supreme Court decisions ruled that citizens are not subject to the same constitutional standards in conducting searches of persons as are the police. This is because the Constitution is designed to protect citizens from government, not citizens against citizens.

Laws are classified as either private or public. *Private law* deals with relationships between individuals in which government does not have a direct interest. Included in this group are laws that address marriage, divorce, contracts, real estate, and personal inquiry. Public laws, on the other hand, are more directly concerned with government interests. Some examples of public laws are criminal laws and criminal procedure.

Public laws also deal with health, safety and public welfare. Some of these laws are called administrative laws because they require businesses or agencies to undertake care in providing public services. Restaurants and liquor establishments must assure that food and beverages are properly prepared and handled. Nightclubs must avoid serving liquor to minors. In general, laws come from state or federal court decisions, the United States Congress, state legislatures, or from local ordinances. Laws pertaining to alcohol and beverage control are enacted by state legislatures.

CRIMINAL LAW AND PROCEDURE

Criminal laws reflect the moral and ethical beliefs of society. Crimes are generally classified as either felonies or misdemeanors. A *felony* is more serious than a misdemeanor. A person convicted of a felony could be sentenced to state prison or receive the death penalty. A felony conviction on a person's record could seriously hinder his or her chances of entering certain professional fields (for example, becoming a police officer). A person convicted of a felony may be prohibited from carrying firearms or holding certain public offices. There are a number of crimes which are grouped as felonies. All state laws define murder, manslaughter, robbery, burglary, and arson as felonies. Murder is considered a capital offense, punishable by death in some states.

Misdemeanors are less serious than felonies, and usually result in fines, probation, or jail terms of usually less than one year. Conviction of petty theft and trespassing are examples of misdemeanors. Conviction of traffic offenses are considered misdemeanors in some states, while in others (for example, California) traffic offenses are considered infractions. It must be understood that states classify felonies and misdemeanors differently. In some states, there are varying degrees of felonies and misdemeanors, each carrying different punishment; therefore, it is recommended that you review your state criminal code to better understand criminal definitions and punishments.

A *crime* is a violation of state or federal public law. In other words, each state has a criminal code as does the federal government. A crime or public offense must have a criminal act (*actus reus*) and criminal intent (*mens rea*). The criminal act can involve intentional omissions or commissions. A *crime of omission* is the failing on the part of the person to perform an act required by law. Failing to file income taxes would be a crime of omission. A *crime of commission* is more direct. For example, a person who shoplifts, assaults another, or burglarizes a business has committed a crime of commission. The law does not hold accountable those persons who are unable to voluntarily commit an act. The American Law Institute's Model Penal Code defines the nature of a voluntary act for criminal purposes:

Section 2.01 *Requirement of Voluntary Act; Admission as Basis of Liability; Possession is an act.*

(1) A person is not guilty of an offense unless his liability is based on conduct which includes a voluntary act or the omission to perform an act of which he is physically capable.

(2) The following are not voluntary acts within the meaning of this section.
    (a) a reflex or convulsion;
    (b) a bodily movement during unconsciousness or sleep;
    (c) conduct during hypnosis or resulting from hypnotic suggestion;
    (d) a bodily movement that otherwise is not the product of the effort or determination of the actor, either conscious or habitual.

(3) Liability for the commission of an offense may not be based on an omission unaccompanied by action unless;
    (a) the omission is expressly made sufficient by the law defining the offense; or
    (b) the duty to perform the omitted act is otherwise imposed by law.[1]

Some acts committed by accident, mistake, or in self defense can be a defense to a crime. For example, a person who accidentally bumps into somebody in a crowded room has not committed an unlawful act even if an injury results.

The criminal law requires proof beyond a *reasonable doubt* that the guilty act was intentional. In other words, to shoot another with the intention of harming the person (causing serious injury or death) would be *criminal intent*. The law refers to criminal intent as *malice* or guilty mind. It is often difficult to determine a person's state of mind; however, there are a number of states that define *criminal intent* or *criminal culpability*. The following section from the Model Penal Code encompasses some of these states of mind. As indicated, *knowledge* is a more serious state than recklessness or negligence. An intruder entering private property carrying a concealed firearm for the purpose of committing a robbery would be displaying criminal intent (knowledge). Playing with a loaded firearm resulting in an accidental discharge would be considered *recklessness* if an injury resulted. Leaving a loaded firearm unattended would be considered *negligence* if an injury resulted from its discharge.

Section 2.02 *General Requirements of Culpability*

(1) Minimum requirements of culpability. Except as provided in Section 2.05 a person is not guilty of an offense unless he acted purposely, knowingly, recklessly or negligently as the law may require with respect to each material element of the offense.

(2) Kinds of culpability defined. A person acts purposely with respect to a basic element of an offense when:
    (a) if the basic element involves the nature of his conduct or a result thereof, it is his conscious object to engage in conduct…

      (b) if the element involves the attendant circumstances he knows the existence of such circumstances.

(3) Knowingly. The person acts knowingly with respect to the basic elements of offense:
      (a) if the element involved the nature of his conduct or the attendant circumstances he knows that his conduct is of that nature or he knows of the existence of such circumstances; and
      (b) if the element involves the result of his conduct he knows that his conduct will necessarily cause such result.

(4) Recklessly. A person acts recklessly with respect to a material element of an offense when he consciously disregards a substantial and unjustifiable risk that the material exists or will result from his conduct...

(5) Negligently. A person acts negligently with respect to a material element to an offense when he should be aware of a substantial and unjustifiable risk that the material element exists or will result from his conduct...[2]

Crimes can be committed against a person or property; moreover, the use of electronic surveillance systems and other equipment presents an opportunity for security professionals to witness crimes. Thus, many criminal acts are being recorded. It is important for the protection officer to understand the different types of crimes against person and property. Following is a brief distinction between crimes against person or property.

Crimes Against the Person

Crimes against the person involve some type of physical assault or injury to another. There are several types of crimes that can be committed against a person.

1. *Murder.* The willful premeditated killing of a human being by another. This does not include deaths caused by accident, or negligence or self defense.[3] Example: A spouse hires a hitman to kill the other spouse. This hitman lies in wait for the spouse and fatally wounds him (the hitman has committed murder as well as the spouse who hired the hitman).

2. *Robbery.* The taking of personal property from another by using force or fear or threat of force.[4] Example: A robber threatens to kill a victim unless the victim gives the robber money. Robbery does not require a weapon.

3. *Assault.* An assault is an unlawful attempt to cause harm with an apparent present ability to harm another. There is no physical contact. An assault is termed aggravated if one attempts to cause the serious physical harm to another with a deadly weapon.[5] Example: An argument between two people results in one attempting to hit the other with a club.

4. *Battery.* Battery is an unlawful touching or physical contact of another. For example, an unlawful push or grabbing would constitute a battery. Many state laws group assault and battery together. A battery is the completion of an assault.

5. *Kidnapping.* A person is guilty of kidnapping if he unlawfully removes another from his/her place of business, residence or other area.[6] For example, to take a person against his/her will to a specific place is kidnapping.

6. *Rape.* Rape is generally defined as the unlawful sexual intercourse of another. Rape is usually accomplished through the use of force, fear, or threat of force but can be accomplished in some jurisdictions by the use of instruments.[7]

7. *Indecent Exposure.* One who unlawfully or with the intent to arouse exposes his/her private parts in a public place is guilty of indecent exposure.[8]

Security personnel will most likely encounter crimes against property. Since one of the primary concerns of security is property protection, a discussion of several types of property offenses are presented.

1. *Arson.* The unlawful starting of a fire or causing a fire for the purposes of property destruction or to collect insurance.[9] Arson sometimes occurs in conjunction with other crimes (for example, murder).

2. *Burglary.* The unlawful breaking or entering of a structure with the purpose of committing a theft or other felony.[10] An intruder who unlawfully enters a building for the purpose of removing property has committed burglary. If it can be proved that a person entered a store with the intent to shoplift, a charge of burglary may also be warranted. The issue in burglary is the offender's prior intent. Also consider that the crime of burglary does not have to be a breaking. Unlawful entry can be accomplished for example, through the use of a passkey or entering through an unlocked door (also known as *constructive breaking*).

3. *Theft or Larceny.* One who unlawfully takes and carries away the personal property of another with felonious intent, is guilty of larceny (intent is key). Depending on the value of the property, a larceny can be either petty or grand larceny. Shoplifting is a common example of larceny.[11]

4. *Vandalism or Malicious Destruction of Property.* One who unlawfully defaces, destroys, or causes damage to property is guilty of a crime of vandalism or malicious destruction of property. Spray painting buildings is an example of vandalism.

5. *Embezzlement.* One who is employed by a business and during the course of employment unlawfully takes money or property belonging to the business (i.e., a bank teller who unlawfully takes money from the till) is guilty of embezzlement.

Public Order Offenses

A protection officer may encounter offenses associated with public order and decency.

Disorderly Conduct. A person is guilty of disorderly conduct if, with intent to cause public inconveniences, that person makes unreasonable noise, fights, or uses obscene language. An officer assigned to a movie theater, library, or other public setting may encounter such offenders. For example, a person who is shouting obscenities and attempting to start fights with other patrons is disorderly and may be escorted from the property.

Public Drunkenness. A person is guilty of this offense if he/she appears in a public place under the influence of alcohol to the degree that he/she may endanger him or herself or other persons or property. These types of disturbances may be encountered in a number of settings, especially parking garages, parking lots, transportation centers, and liquor establishments.

There are various stages involved in criminal procedure depending upon how the case is processed. Once a crime has been reported, an arrest may occur. An *arrest* involves restricting a person's freedom and taking that person into custody for possible prosecution. Before the trial, there are pretrial hearings to determine if there is enough evidence to hold someone for trial. As indicated in Figure 2–1, an arrest can result in an acquittal or conviction. In many cases,

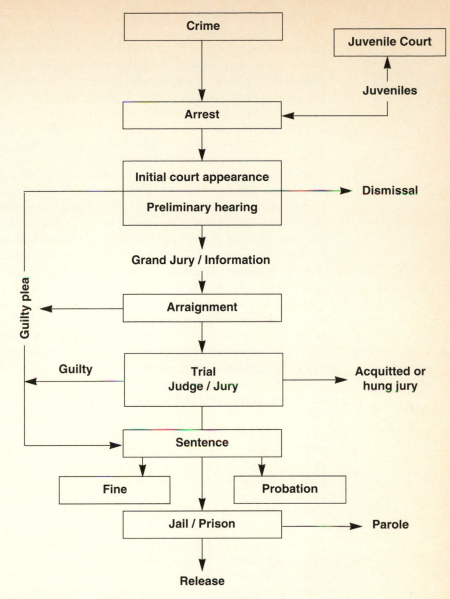

```
                        ┌──────────────┐
                        │    Crime     │              ┌────────────────┐
                        └──────┬───────┘              │ Juvenile Court │
                               │                      └────────▲───────┘
                               ▼                               │
                        ┌──────────────┐              Juveniles│
                        │    Arrest    │◄─────────────────────
                        └──────┬───────┘
                               │
                               ▼
                  ┌─────────────────────────┐
                  │ Initial court appearance │─────────►  Dismissal
                  │                          │
                  │   Preliminary hearing    │
                  └────────────┬─────────────┘
                               │
                        Grand Jury / Information
                               │
                  ┌────────────▼─────────────┐
          ◄───────│       Arraignment        │
                  └────────────┬─────────────┘
                               │
          Guilty   ┌───────────▼─────────────┐
          ◄────────│          Trial          │──────►  Acquitted or
                   │       Judge / Jury       │         hung jury
                   └───────────┬─────────────┘
                               │
                  ┌────────────▼─────────────┐
          ───────►│        Sentence          │
                  └──┬──────────────────┬────┘
                     │                  │
          ┌──────────▼───┐      ┌───────▼──────┐
          │     Fine     │      │  Probation   │
          └──────────────┘      └──────────────┘
                     │
                  ┌──▼───────────────────────┐
                  │       Jail / Prison       │──────►  Parole
                  └────────────┬─────────────┘
                               │
                               ▼
                           Release
```

FIGURE 2–1   The flow of criminal cases.

offenders will plea bargain or plead guilty to a lesser charge in order to get a reduced sentence. Since our justice system is clogged with cases, this bargaining process is often necessary. If there is a trial, then the person can be found guilty or acquitted. If a verdict of guilty is found, then the defendant is sentenced. Sentencing can be anything from a fine to death, depending upon the seriousness of the offense and the offender's prior record.

Arrest and Detention

The power of citizens to make arrests is derived from common law and has been preserved by every state by civil law or statutory authority. Private-person arrest powers are not the subject of the Constitution. In other words, the Constitution does not apply to private persons or protection officers acting on their own. An arrest can be made with or without a warrant. However, as a protection officer, you will not be making arrests with warrants. In fact, a protection officer won't be making many arrests at all. However, if an arrest is made, it will generally be for some public offense observed by the officer. The best advice is to observe the offense personally rather than taking someone else's word for it. If someone informs an officer that he or she saw someone commit a theft, that alone may not be enough reason for detaining that person. However, it would be reasonable to assist another who observed the crime assuming, of course, that the person observed the offense. Having knowledge of a crime being committed or one that has been committed is *probable cause*.

A private protection officer's arrest authority is generally no greater than any other private person. Unless you are *deputized, commissioned*, or provided *arrest authority* by municipal ordinances or law, your powers are limited in the same way as other private persons. An arrest or detention is a restraint of a person's movement for some lawful purpose. It is not necessary that the person be handcuffed or confined. Simply telling someone to "sit and don't move" may be sufficient to be an arrest. However, there is often some confusion as to the difference between an arrest and a detention. While both involve restraints, an arrest is considered more custodial. Both need probable cause. If a false detention/arrest occurs, a charge may be filed by the "arrestee" that the protection officer committed the *intentional tort* of *false imprisonment*.

When the police arrest an offender, it may involve booking and prosecution. A *detention* is more temporary. It is considered a prearrest. A protection officer who detains an offender may have the offender taken into custody by the police, or held until security reports are completed. In California, as with a number of other states, a private person and protection officer can arrest for misdemeanors committed in their presence or felonies *not* committed in their presence as long as there has been a felony committed. In some cases this means that the officer does not actually have to see the offense before an arrest is made. An example of this would be the situation where a woman reports to a protection officer that she was robbed of her purse. Shortly thereafter, the described offender is seen running from the location. An arrest can be made based upon the reasonable belief that the person running committed a crime. The officer did not actually see the crime, but had reasonable grounds to believe it happened based upon such factors as the victim's description and the offender's flight.

If a protection officer observes a suspicious person loitering in a parking lot, the officer can approach the person and ask if he or she needs assistance, or otherwise invite the person to leave. Whether a suspect is arrested or detained does not matter. What is important to remember is that if a person is being held involuntarily, the officer must have knowledge that a crime has been committed. A protection officer may approach a suspicious person and ask if he or she is willing to answer questions or explain why they are on private property. Before questioning another person, the officer must let that person know that he or she is free to leave. If the protection officer decides to detain someone against his or her will, there must be cause to believe that the person was involved in a crime, or otherwise violated company policy .

The protection officer may assist a person who observed a crime, but only if convinced that the person who saw the offense is positive. If a person must be restrained, the protection officer should use only that amount of force necessary to keep the person under control until the police arrive. This means the protection officer could handcuff the person or use other reasonable physical measures to control the situation. The question is often asked: must a protection officer advise an arrestee of his or her *Miranda* rights*? When the police initiate an arrest, they

---

*The Miranda Ruling requires the police to advise arrestees of their right to remain silent, their right to an attorney, and their right to be provided an attorney if they are unable to retain one on their own.

must advise the arrestee of his or her rights before asking any incriminating questions. However, the protection officer is not a police officer and the law does not require the protection officer to advise persons detained of their rights. Some protection officers give Miranda warnings, but this is done as a matter of their company policy, not a requirement of law (unless the protection officer has police authority). Generally, protection officers should not ask detained persons for anything beyond basic information such as name and address.

In initiating a private-person arrest or detention, the following suggestions are offered:

1. Make sure you have probable cause. This means that you must have seen an offense committed by a particular person.

2. Attempt to get assistance or a witness before initiating a stop. This will help protect you from claims of misconduct.

3. Do not accuse or otherwise verbally abuse the offender.

4. Do not use any physical force unless it is absolutely necessary, and only for self protection. You may use limited force to escort the person to an office.

5. If the person becomes violent or combative, get assistance. If you cannot get assistance, it is better to let the offender go than risk a confrontation which may cause unnecessary injury to you or the offender. If the offender is attacking you, you may use reasonable force to defend yourself.

6. Be polite. Use words such as *sir* or *miss* and phrases such as *I would like to talk with you*. Avoid rough language such as *hey you* or *get over here*.

7. Do not detain someone for any unreasonable length of time. Limit your detention to one hour. This time should give you ample opportunity to complete reports. If your intent is to have the police take the person into custody, notify the police immediately!

8. During detention, have a witness present. If it is a female detainee, a female witness is necessary.

9. Do not forcibly search a person, ask for consent first. This request is done even if you know that the person has merchandise concealed on his person. However, if the offender is in custody and under arrest, a search can be done. Be sure to check local laws.

10. Do not chase a person long distances or into unfamiliar places. This usually means that if you cannot catch or apprehend the person on company property, let him or her go. You can always use your radio to get assistance.

11. Remove the person from public view as soon as possible. Avoid or limit public exposure.

12. When completing reports, be accurate and record any injuries or possible injuries of the person detained.

13. If you think that the person is carrying a deadly weapon, you may search that person for purposes of self protection. This type of search is limited to outer clothing such as pockets.

14. Ask the arrestee to sign a statement indicating that no force or threats were used.

15. If the arrestee confesses to the offense, make an attempt to have the arrestee write the confession.

Suggestions For Handling Shoplifting Cases

For those protection officers employed in retail security, or who work undercover assignments, the following guidelines are offered:

1. Make sure you have probable cause before you restrain the freedom of movement of a person for shoplifting. Probable cause (or reasonable grounds to believe) must be based on personal knowledge (firsthand information) by you or another reliable adult employee. Remember: If you did not see it, it did not happen. When in doubt, do not detain!

2. Observing a person concealing "something" or putting "something" in their pocket or purse is not sufficient to establish probable cause. You must have reasonable grounds to believe (probable cause) that the object is merchandise that has not been purchased and that the item belongs to the store. If you do not have probable cause, you may
   - Keep the person under observation.
   - Engage in voluntary conversation ("May I help you?" "Are you looking for something?").
   - Ask the person whether he or she has a receipt for the merchandise under circumstances in which there is no restraint of the person's freedom of movement.

3. After a person is observed shoplifting, the following options are available to a protection officer:
   - Confront the person immediately and ask him or her to produce the item. Always ask for a receipt showing that they paid for the item. Under these circumstances, you may seek only recovery of the item rather than prosecution in court.
   - You may be under instructions to allow the person to go beyond the last pay station (or in some cases, even out of the store). Under these circumstances, the person should be kept under surveillance. If the person becomes aware of your surveillance, he or she may attempt to discard the shoplifted item or pass it onto another person. If you fail to observe the discard or pass-on, you may then be unable to explain why the stolen property was not recovered. If it appears that the person may outrun you and other store employees, it may be wise to position yourself between the exit and the person.

4. Shoplifting cases are handled by your local law enforcement agency (police or sheriff), prosecutor, or judge. Consult a knowledgeable official to determine
   - whether the local judge requires that the person observed shoplifting be allowed beyond the pay station (or out of the store) before they are detained. (A young person in gym shoes who gets near a door or out of the store is going to outrun most store security people.)
   - the factors affecting prosecution: value of the merchandise stolen, recovery of the merchandise, and the age of the offender.

5. Absolute defenses to civil suits are
   - that the person either voluntarily stayed in the area, or that the restraint of movement was made in good faith on probable cause based on personal knowledge.
   - that if any force was used, it was necessary either in self-defense, or to detain the person, and/or to prevent the unlawful theft of the property, and that the amount of force was reasonable under the circumstances[12]

Search and Seizure

The Constitution is designed to protect citizens from government intrusions rather than citizens from other citizens. However, protection officers acting as agents of police officers (government officers) could come under constitutional guidelines. For example, if a police officer enlists the aid of a protection officer to conduct a search on behalf of the police officer, the protection officer would be acting under state authority. If the search was without probable cause or otherwise unreasonable, the evidence would not be admissible even though the police officer wasn't directly involved in the seizure.

Most states have enacted laws authorizing merchants or their employees to arrest or detain a person suspected of shoplifting. Evidence seized by private protection officers in these situations are generally not subject to the exclusionary rule. *The exclusionary rule applies to public law enforcement officers.* Simply stated, the rule requires that any evidence seized in an inappropriate manner by those working for a government agency (for example, police) cannot be used against the defendant in court. This means that the police must have probable cause in order to search. The rule does not apply to private persons and protection officers. The United States Supreme Court has ruled that evidence seized unreasonably by a private person can be used in court to convict a defendant. Private protection officers are considered private persons. However, this is not to suggest that protection officers search the property of another at random or without reason. A protection officer could be subject to civil suit if a search caused personal injury or property damage. There is a difference between *search* and *seizure*. A *search* requires a looking into a place normally considered private (for example, trunk of car, locker, lunch box). A protection officer who takes contraband from an employee's locker can turn it over to the police. This search and seizure would be the result of some cause on the part of the officer. Protection officers may be allowed to search employees. The policies of some companies permit a search as a condition of employment. The police, on the other hand, would need a search warrant unless the search was by consent or incidental to a lawful arrest.

A *seizure* is the taking of property from another which is the result of plain view. There can be a seizure without a search. For example, if an officer working the main gate of a parking lot observes a firearm in plain view (lying on the seat of a car), the officer could seize the weapon because it was in plain view. If the door was locked, the officer would probably get a description of the vehicle and report it immediately.

The exclusionary rule would apply if it could be proved that the police and citizens were working together. For example, if you conduct a search on behalf of a police officer, your immunity would not apply because you have become an agent of the state or a police assistant. As a professional protection officer, be on guard that you are not tricked or coaxed into acting as an agent of the police. Remember, if you detain or otherwise restrict someone's movement (for example, a shoplifter or trespasser), you are also seizing the person; thus, the reasonableness of that seizure would depend on your reason or probable cause for detaining in the first place. A protection officer's authority to detain or seize another also depends upon company policy, type of business, and the intruder's intentions. Generally, trespassers should be told they are trespassing and given an opportunity to leave before detaining or making an arrest. Avoid detaining or seizing someone if alternative courses of action exist.

CIVIL LAW AND PROCEDURE

A civil wrong is a private wrong. Contract violations, business disputes, and personal injuries are examples of civil wrongs. In civil cases, as with criminal cases, there are certain procedures that must be followed. In civil suits the primary aim is to seek monetary awards rather than criminal prosecution. Civil cases begin with a complaint filed by a *plaintiff* (the person who is injured) against the *defendant* (person or persons who are allegedly responsible for the injury or harm). The defendant will then file an answer or response to the complaint which usually disputes the plaintiff's claim. There will be a series of pretrial activities which include *depositions* (out of court testimony) and hearings. Most civil cases are either settled out of court or dismissed, with a few going to trial (Figure 2–2). Parties to lawsuits may be required to testify regarding knowledge of or involvement in the incident. If a *judgment* is granted, the defendant pays damages to the plaintiff.

Civil suits can also result from criminal acts. A person who commits battery on another can be charged with a criminal offense and can also be sued for causing physical or emotional damage. For example, if the victim receives a broken arm from the attack, the offender may be required to pay medical damages if a civil judgment is found against him or her. There are courts for criminal cases and civil cases. In civil court, there is no need to prove beyond a reasonable doubt. All that is needed is a preponderance of evidence to show that a person (plaintiff) was harmed.

LIABILITY

In the security business, you and your employer will be threatened constantly with law suits. We live in a litigious society; therefore, it is crucial that you per-

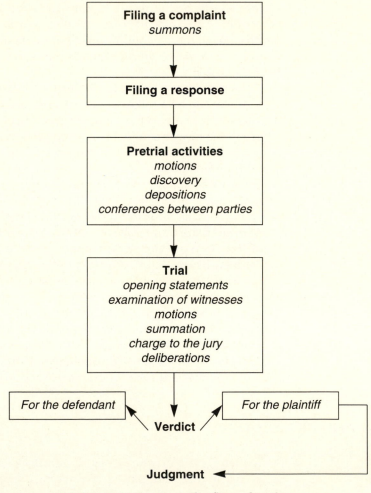

FIGURE 2 – 2   The flow of civil cases.

form your job carefully and reasonably. Most cases involving security litigation occur in apartment or condominium complexes, retail stores, hotels, motels, shopping malls, office buildings, parking lots, and other public areas. Law suits directed at private security frequently result from 1) negligence or torts committed by the security company or its employees, or 2) criminal acts committed by the security company or its employees (Cunningham and others, 1990, p. 35). A tort is a civil law term referring to an injury caused by the negligence of another. A *tort* is defined as a civil wrong, other than a breach of contract, in which the action of one person causes injury to the person or property of another in violation of a legal duty imposed by law. In security, the two most frequently encountered types of torts are the intentional tort and the negligent tort.

Intentional Torts

Intentional torts can be committed against a person or property. An *intentional tort* occurs when there is an intention on the part of a person to bring some physical or mental harm upon another person. Intentional torts relevant to security duties are 1) excessive use of force, 2) false arrest or imprisonment, 3) assault and battery, 4) wrongful death, and 5) slander.

Excessive Force. *Excessive force* occurs when the force used is unreasonable for the situation. Thus, a protection officer attempting to detain a shoplifter by using a chokehold which causes injury to the shoplifter would be guilty of using excessive force. Even if the shoplifter attempted to flee, the use of extreme levels to control or detain would be viewed as excessive. Protection officers who carry batons, handcuffs, or chemical agents must be careful not to use excessive force. Any form of touching which results in injury could be subject to a lawsuit. In civil suits alleging use of excessive force, the plaintiff or injured party will attempt to show that the protection officer failed to use alternatives to force. Some alternatives to force are using verbal persuasion, calling for assistance, or simply letting the person flee.

False Arrest. In *false arrest* cases, there is an allegation that an officer made an illegal arrest or detention under state law. This usually means that an arrest occurred without probable cause. Charges of false arrest can arise in shoplifting cases where protection officers stop someone for theft without actually having seen the offense and no merchandise is found on the suspected offender. *False imprisonment* usually follows false arrest in that it involves a detention without probable cause. A protection officer may make a valid arrest, but hold the person for an unreasonable length of time. On the other hand, if the protection officer initiates a false arrest, then the continued detention would be false imprisonment. A good rule of thumb is to never detain anyone longer than thirty minutes to an hour. If detaining a juvenile, the protection officer may choose to release the juvenile to a parent or guardian as opposed to the police. It may take time for the parent to arrive, but at least the proper notifications have been made.

Assault and Battery. An assault and battery upon another can result in a tort action against the protection officer. A protection officer could be sued, if, while carrying a baton or club, he or she strikes another without justification. If the unjustified striking results in the death of another, then a *wrongful death* lawsuit may be filed by the surviving family, relatives or legal guardians. As discussed previously in this chapter, the protection officer could be charged criminally for causing serious injury and then be subject to a civil suit arising out of the criminal act.

Slander. Slander can be the source of an intentional tort. Basically, *slander* is an oral accusation against another in public. It is a statement defaming the character of another. For example, if a protection officer publicly accuses another of being a thief (and it is untrue), that is slander. If, however, it is learned that the accused person is a convicted thief, that information could be used as a defense to slander.

Negligent Torts

Another area of civil law relevant to security is general negligence. *Negligence* is the absence of care, according to the circumstances, causing injury to another. Negligence occurs when a protection officer, acting unreasonably or carelessly, harms another even though the injury may have been unintentional. In order to establish negligence, it must be shown that the defendant protection officer had a duty to protect, and the protection officer failed to exercise that duty. In other words, the *determination of negligence* involves: 1) a duty to protect, 2) a violation or breach of that duty, 3) a relationship between a duty to protect and breach of duty, and 4) there must be injury (damage). Related to this definition is foreseeability or the knowledge that the incident was predictable or likely to happen due to some security problem.

General negligence differs from an intentional tort in that the former implies carelessness on the part of the officer, while the latter is more intentional or direct. For example, an officer who falls asleep at his post would be acting negligently if an intruder slipped past and caused injury to another. The protection officer had a duty to protect (control access) but violated (breached) that duty by not staying awake. The injury received by the third party resulted from the guard falling asleep at his post.

Negligence results from inattention to duty or the failure to perform an agreed-upon task. It is the protection officer's responsibility to follow post orders and be attentive. The protection officer's failure to perform a required task or uphold company policy can result in the officer and his or her employer being charged with negligence should third-party injury or harm result. In other words, if the post assignment requires making rounds every hour and checking specific locations, then the protection officer is expected to do so. If the protection officer neglects that responsibility, both the protection officer and his or her employer could be held negligent if an injury or harm befalls another. Under negligence theory, the issue is not whether you intended to harm another, but whether your failure to perform a required duty caused injury to another. The point is that the protection officer or the employer should know better. Simply saying, "I forgot," or "I didn't mean to cause harm," is not enough to defend against civil liability.

An intentional tort could arise if a protection officer observes an intruder and inflicts excessive force in order to "teach" the intruder a lesson. A professional protection officer must recognize the duty to protect and the importance of acting reasonably in the performance of that duty. This duty does not suggest that you sacrifice your life to save others, but that you take steps to assist those in danger. If, for example, you observe an armed attack in your assigned area of protection, you would be expected to call for assistance or notify others of the activity. You would be expected to use your wits to distract or discourage the attacker as much as possible. The simple act of blowing a whistle or shouting at the intruder could be enough to deter the attack. The protection officer is not expected to be a human shield for another. If armed with a firearm, then appropriate shoot-don't-shoot measures would apply (see Chapter 7). This, of course, would depend upon the protection officer's training, the company use-of-force policy, state laws, as well as the seriousness of the attack.

In security, lawsuits are often difficult to avoid; yet, the threat of a lawsuit should not be a deterrent from performing your duties. In other words, acting reasonably and responsibly increases the chances of withstanding a lawsuit.

Sexual Harassment in the Work Place

The modern protection officer is confronted with a number of legal challenges. There may be situations when an officer may be informed of harassment incidents, or be named a party to a harassment charge. The officer should be familiar with the issues and legal implications of sexual harassment claims in the work place.

Sexual harassment is a form of sex discrimination that violates federal law under what is as known as Title VII of the Civil Rights Act of 1964. Unwelcome sexual advances, requests for sexual favors, and other verbal or physical conduct of a sexual nature constitute *sexual harassment* when agreeing to or rejection of this unwelcomed conduct affects an individual's employment; unreasonably interferes with an individual's work performance; or creates an intimidating, hostile, or offensive work environment.

Sexual harassment can occur in a number of circumstances. The following is list of examples:

- The victim as well as the harasser may be of either sex. The victim does not have to be of the opposite sex.
- The harasser can be a supervisor, a co-worker, a supervisor in another department, or a nonemployee.
- The victim does not have to be the person harassed but could be anyone affected by the offensive conduct such as a third party overhearing or seeing the conduct.
- The harasser's conduct must be unwelcome.
- There need not be any economic or physical injury to the victim in order for a complaint to be made.

If an individual feels that he or she has become the victim of sexual harassment, he or she should inform the harasser that the conduct must stop. Normally, businesses will have complaint procedures or grievance systems available. However, charges of sexual harassment may be filed at any office of the U.S. Equal Employment Opportunity Commission. These offices are located in fifty cities throughout the United States.

Anyone discriminated against due to sexual orientation may be entitled to a remedy. These remedies may include promotion, reinstatement, and back pay. A victim may also be entitled to damages to compensate for mental anguish.

To illustrate the seriousness of sexual harassment incidents, the following actual case is presented:

A single, twenty-five-year-old mother of two children brought suit against her employer. She complained that she had been sexually harassed while on the job. Evidence indicated that the harasser made "repeated offensive sexual flirtations, advances, propositions, and sexually offensive conduct." She reported this to her supervisors, but they took no action against the harasser. The harasser was also her supervisor and gave her extra work when she refused his advances. A jury returned a verdict awarding $45,000 in actual damages and $45,000 in punitive damages to the employee for her emotional distress. The decision was decided by a federal court (*Baker* v. *Weyerhaeuser Co.*, 903 F. 2nd, 1342, 1990).

As with the protection philosophy, prevention is the best method to avoid claims of sexual harassment. Employers have the responsibility to inform workers of sexual harassment laws and claim procedures. Employees have the responsibility to treat each other with respect in the work place. Due to their responsibility of upholding work place rules, protection officers must particularly avoid charges of sexual harassment.

The following guidelines are offered as a means to reduce claims of sexual harassment in the work place:

1. Avoid physical contact with co-workers or invitees unless assisting or restraining someone in a manner authorized by law or company procedure (medical assistance, arrests).
2. Do not display sexually oriented material (for example, posters, magazines) in the work place.
3. Avoid jokes or other verbal exchanges that contain sexual language that may be offensive. Sexual references such as "nice legs" or "good body" while offered as compliments, may still be offensive to some.
4. Excessive staring should be avoided unless associated with the protection role. In other words, harassment claims may be initiated against a person who engages in inappropriate and prolonged eye contact. Staring is frequently accompanied by suggestive sexual comments, whistling, or nonverbal gestures.
5. Do not question or otherwise discuss a person's sexual past or preferences.
6. Do not initiate nonverbal gestures which suggest sexual encounters or behaviors.
7. Do not offer promotions, choice assignments, preferred working schedules, or other favorable conditions in return for sexual favors. Such offers are definite grounds for sexual harassment claims.
8. If seeking a relationship or date with another, and that person refuses, do not continue to ask. In short, no means no.

Claims of sexual harassment can be initiated against a member of the same sex as well as the opposite sex. Harassment charges can occur even if the offensive language was directed at an individual other than the victim. An example would be a group of male officers exchanging offensive sexist jokes in the presence of female officers, even though the female officers were not the targets of the comments.

Closely associated with sexual harassment claims are situations involving racial harassment. As with sexual harassment, inappropriate comments regarding a person's race or ethnic background can be a cause of action under federal law. Protection officials are composed of persons of many races and nationalities. Therefore, unwelcome jokes, gestures, or written materials insulting to another's heritage must be avoided. A professional protection officer must maintain an image that commands respect and cooperation. This image translates not only to personal appearance, but to language and behavior as well.

Respondeat Superior Liability

There are other sources of liability confronting the security industry. The following types pertain to mismanagement. As a professional protection officer or supervisor, it is important for you to recognize how your company could be liable under the category of negligence called *respondeat superior liability* or *vicarious liability*. It is possible for management to be held responsible for the unlawful acts of employees. In other words, an employer can be held liable for the negligent acts of subordinates if it can be proven that management failed certain supervisory responsibilities. This is referred to as vicarious liability.

Negligent Hiring. This occurs when an agency or company fails to perform adequate or reasonable preemployment screening or background checks of prospective employees. Management has a duty to assure that employees assigned to sensitive protection responsibilities are trustworthy. This type of negligence is often the result of a failure on the part of management to check an applicant's references or criminal and credit history.

Negligent Training. This type of liability is the direct result of inadequate employee training. Protection officers who are negligent in their duties may not have had the proper training; thus, management would be considered negligent. This type of negligence occurs when officers are given a post assignment or a weapon to carry without proper training.

Negligent Retention. This area of management liability results from management's failure to dismiss an officer when it is known that the officer's performance is detrimental to the agency. For example, a company that ignores client complaints of excessive use of force by one of its officers could be charged with negligent retention of an unfit employee.

Negligent Supervision. This liability usually results from improper supervision or policy guidance. For example, supervisors are responsible for assuring that subordinates are properly carrying out their duties according to policy. Also, supervisors must understand policy and the law relevant to their duties. A supervisor who gives an improper order that results in injury could make the employed vulnerable to charges of negligent supervision.

Criminal Liability

Criminal acts committed by security employees while on duty can result in liability. If a protection officer burglarizes a business which he is responsible for protecting, or unlawfully injures a client, the protection officer can be held criminally responsible. As previously discussed, a crime or public offense is an unlawful commission or omission coupled with criminal intent completed against another. For example, a protection officer protecting a parking lot who unlawfully removes the personal property of another is committing a theft and could be prosecuted under appropriate state law. A protection officer assigned to an apartment complex who uses a passkey to unlawfully enter an apartment for the purpose of committing a sexual assault or burglary, could be charged with a criminal offense.

These aforementioned criminal acts would obviously lie outside the contractual duties of a protection officer. Security companies do not hire protection officers to commit crimes; therefore, the company would not necessarily be responsible for such criminal actions unless it could be proved that the employer failed to check an officer's background or provide proper supervision. However, an employer can usually be held liable for the actions of a subordinate under the theory of respondeat superior or vicarious liability. When a crime results in injury (emotional or physical), the offender has committed a public offense as well as a civil offense. Although intentional torts are generally classified as civil offenses, an intentional tort can be considered criminal when the act is excessive. For example, the use of excessive force to subdue an intruder (causing injury) would normally be considered an intentional tort; however, if the force escalated to a severe beating (as in the infamous Rodney King case in Los Angeles), criminal charges could be filed against the protection officer. An intentional malicious beating could result in criminal charges being filed against the protection officer. To cite another example, in some states, slander is a criminal offense as well as a negligent tort. As previously discussed, slander is simply the act of defaming another person's character through oral (public) accusations. A protection officer who falsely arrests another for shoplifting, and subsequently publicly accuses the person of being a thief has committed slander as well as false arrest. The officer could be charged under some state slander criminal codes and face subsequent civil action for defamation of character.

Guidelines For Avoiding Civil and Criminal Liability

The job of a protection officer is not easy. There are many temptations and litigation threats. Unfortunately, liability cannot be avoided altogether. The following guidelines are provided to assist the protection officer in performing his or her job and to reduce the chances of liability. Remember, if you are unsure about what to do in any given situation, consult a supervisor. It is the employer's responsibility to make sure the protection officer understands assigned duties and it is the responsibility of the protection officer to properly perform those duties.

1. Do not touch, grab, or use any physical contact on another unless absolutely necessary.
2. Do not draw or display a weapon unless intending to use it for purposes of self defense or defense of another (see number 5).
3. Do not violate or otherwise deviate from post orders or assignments without approval. This also means remaining at the post until properly relieved.
4. If physical force must be used, use only the minimum amount of force necessary to overcome resistance. This means that force must only be used as a last resort for purposes of self defense where retreat and/or assistance is impossible.
5. Never use chokeholds or carotid restraints.
6. Never use head strikes with a baton.
7. Never fire warning shots.
8. Never handcuff too tightly. Assure that there is sufficient circulation and that the cuffs are double locked.
9. Do not watch television, listen to radios, or read while on duty.
10. Do not socialize with clients, other employees, or invitees while on duty.
11. Do not use alcohol or drugs while on duty.
12. Make sure your equipment (radios, vehicles) is in good working order before going on duty. If not, report it!
13. Report any unusual incidents and suspicious persons immediately.
14. Detain a shoplifter or intruder only for the brief period of time required to gather needed information.
15. Do not search a person or his or her property unless in their presence, with a witness, and with their consent. If there is no consent, call a supervisor. Do not get into a needless confrontation.
16. Do not use sexist or racist language. This includes avoiding jokes and written material insulting to a particular racial or ethnic group.

17. Operate all vehicles and equipment in accordance with company policy and state laws.

18. Never use any instrument or piece of equipment unless trained in its use.

19. Never detain or otherwise restrain another unless you have personal knowledge that the person has committed an offense. Do not take anyone else's word for it.

20. If detaining another, always treat the person with respect. Do not argue or use offensive language, even if verbally challenged by the offender.

21. In completing reports or activity sheets, make sure to report the facts honestly and accurately. Be specific about times and events. Do not give opinions or include information that did not occur.

22. In cases of injury or illness, immediately call for medical assistance.

Conclusion

This chapter acquaints the protection officer with general legal procedures and issues. Whether the officer is uniformed or employed in a public or private business, the need to understand legal limitations is important. A uniformed protection officer may make fewer arrests than an undercover protection officer; however, the public will often seek assistance from anyone who appears to have some legal authority. This does not mean the protection officer should know as much about the law as a police officer, but he or she should have sufficient knowledge about what not to do. Local laws should be reviewed before assuming any assignment.

Discussion Questions

1. What are the sources of American law?
2. What is the purpose of a three-part government?
3. Compare and contrast crimes against persons and crimes against property.
4. What is a crime of omission as compared to a crime of commission?
5. Define *criminal law* and explain criminal procedure.
6. Define *civil law* and explain civil law procedure.
7. What is a negligent tort? Give an example.
8. What is an intentional tort? Give an example.
9. What is the difference between criminal liability and civil liability?
10. Explain respondeat superior liability. List the four types of respondeat superior liability.
11. When can a protection officer arrest another? Give an example.
12. Under what circumstances can a protection officer conduct a search?
13. Explain how a person can be criminally and civilly liable for the same act .
14. When does the exclusionary rule apply to protection officers?
15. Why aren't protection officers required to give Miranda warnings to persons they detain?
16. Discuss the law regarding sexual harassment.

Discussion Scenarios

How Would You Respond?

Situation 1: The Shoplifter

You are assigned to a plainclothes shoplifting detail at a department store. The store manager informs you that a person removed a jacket from the store without making payment. You subsequently observe the described person hurriedly walking from the store into the parking lot. The manager instructs you to apprehend the offender. Based on the facts given, what would be your response?

Situation 2: Suspicious Person

You are assigned to a fixed-post gate assignment at a parking garage located in an area with a high rate of crime. Entry to the garage is restricted to building employees only. A well dressed, unidentified pedestrian enters the garage, ignoring your demand to stop. The person tells you that he is an employee of the building, but you do not recognize him. How would you respond to the incident?

Situation 3: Possible Theft

A man was seen leaving a men's clothing store with a bulge under his overcoat. A protection officer who had arrested the man for shoplifting on two prior occasions, followed the man to a barber shop. When the man sat down in the barber shop, the protection officer returned to the clothing store, where he determined that a man's leather coat was missing. The protection officer started back to the barber shop and saw the man on the street.[13] What should the officer do?

Situation 4: Unreasonable or Reasonable Force?

A hotel protection officer is informed that two suspicious persons are loitering and using profanity in the hotel parking lot. The two persons appear to be intoxicated. The officer advises them to leave but they refuse and are verbally abusive. They inform the officer that they are guests of the hotel. The officer removes his baton and warns, "guest or no guest, you must leave the lot." The guests warn the officer that if he doesn't leave them alone, they will take his baton away and "break it over your head." The officer strikes one of the persons with the baton causing a compound fracture. Is this reasonable force under the described circumstances?

Situation 5: Search or Not

An employee approaches a protection officer advising him that another employee has a firearm in his locker. Firearms are strictly prohibited on company property. The complaining employee has been with the company a long time and appears to be trustworthy. However, the complaining employee does not want to be involved. The locker is locked, with the only key in the possession of the suspected employee. What should be the officer's response?

Situation 6: Drugs on Campus

A school protection officer is approached by undercover police officers informing him that a student is selling crack cocaine out of his vehicle. The suspected student is known to have a history of pushing drugs on campus. Since the school day is almost over, the police ask the officer to search the student's van. The officer agrees and retrieves ten pounds of cocaine. The officer hands the drugs over to the police. Is this a reasonable search? Would it make a difference if the vehicle was locked? Explain.

Situation 7: The Suspicious Person

While patrolling a shopping-mall parking garage, an officer observes a suspicious person looking into parked vehicles. The mall has had a history of stolen vehicles and vehicle vandalism. The individual appears nervous and begins to walk away when he notices the officer. Based upon this information, what should be the officer's response?

Situation 8: Access Control

An officer is assigned main-gate duties at a large manufacturing plant. The post orders require the guard to make sure all employees entering the plant have picture identification. During a shift change, an employee attempts to enter without ID. He states he forgot his ID. Since it is late at night, attempts to verify his employment with personnel are not possible. The employee is getting angry and says if the guard doesn't let him enter he will "jump all over" him. What should be the guard's response?

References CUNNINGHAM, WILLIAM C., AND OTHERS, *The Hallcrest Report: Private Security Trends 1970–2000*. Stoneham, MA: Butterworth-Heinemann, 1990.

GARDNER, THOMAS J., *Criminal Law Principles and Cases*. St. Paul: West Publishing Company, 1992.

Notes 1. American Law Institute Model Penal Code §2.01 (1962) hereinafter referred to as Model Penal Code.

2. Model Penal Code, *supra* note 2 at §2.02, p. 13.

3. *Id.*, §210.2.

4. *Id.*, §§222.1.

5. *Id.*, §§211.1–211.2.

6. *Id.*, §212.1.

7. *Id.*, §213.1.

8. PA. Statutes Annotated Title 18 §3126.

9. Model Penal Code, *supra* note 2 at §220.1.

10. A TREATISE, *supra* note 1.

11. *Id.*, 802.

12. Gardner, Thomas J., *Criminal Law Principles and Cases* (3rd ed.), p. 327. St. Paul: West Publishing Co., 1985.

13. Ibid, p. 333.

Patrol Procedures

After studying this chapter, you should be able to explain the following:

Conducting preliminary investigations
Preparation for patrol
Crowd control methods
Preserving evidence
Testimony
Radio communications

Patrol equipment
Unusual incidents
Protection officer deportment
Techniques of foot and vehicle patrol
Types of physical evidence
Watchclocks and bar-code systems

INTRODUCTION

The uniformed protection officer is the front line of defense. To many people the uniformed officer represents authority, knowledge, and responsibility. The officer is a representative of the business or facility being protected. The manner in which the officer presents himself or herself can have a lasting effect on the public. There are many factors that affect an officer's preparation for duty. The most important is attitude. It is important for protection officers to maintain a positive attitude in spite of the sometimes difficult people and challenging situations they encounter. Recognizing that only a small percentage of people may behave in a negative manner will assist the protection officer in coping with a variety of experiences. The protection officer's personal value system must be the same as the objectives of the business being protected. Thus, the protection officer must work toward maintaining an attitude of confidence and respect for all people. To counteract negative feelings, an officer should associate with people who have similar values and get involved with hobbies or activities outside of the job. This is important in order to maintain perspective on life and your fellow humans. In short, the way officers treat others, and present themselves in uniform, is the way people, including co-workers, will respond to them.

PREPARATION FOR PATROL

A preparing patrol officer should be armed with information and equipment. Patrol responsibilities may include parking lots, commercial establishments, and private residential tracts.

Information Gathering

There are a number of information sources a protection officer should consult before assuming a new post. A briefing will provide a summary of pertinent information about the post area such as prior instructions and potential safety hazards (for example, lighting problems, broken windows). The protection officer may receive orders to complete assignments begun by an earlier shift.

Reviewing previous patrol logs and written reports will also assist the officer in preparing for the upcoming tour of duty. Previous reports may detail the actions of suspicious employees, unauthorized intrusions, or faulty equipment.

Company policy; local laws and ordinances; business organization, layout and philosophy will all affect the protection officer's duties. Policies restricting employee parking and entrances as well as precise location of offices, buildings, emergency equipment, power and water systems, and storage of hazardous materials should all be familiar to the officer. In policing, an officer must know the beat. This includes geographical boundaries, streets, businesses and other concerns. A protection officer must also know the beat.

Equipment

Preparing for patrol requires an inspection of equipment. An assignment may require the protection officer to carry a two-way radio, a firearm, or operate a vehicle. Whatever the equipment, it must be inspected and maintained prior to going on duty. All signs must be posted properly and provide clear warning to outsiders. The protection officer does not assume anything operates unless it is first checked!

There are some general tools of the trade for protection officers. The following list is offered as a guide to some of the tools required for duty. Naturally, the type of equipment carried will depend upon the assignment and the training.

1. Pocket-size notebook and pen
2. Flashlight (Kel-light suggested)
3. Two-way radio
4. Personal identification including any certificates (for example First Aid, CPR)
5. List of emergency phone numbers or other contacts (written in notebook)
6. Copy of post orders, duties, or other pertinent confirmation of your protection area
7. Watch
8. Handcuffs*
9. Baton*
10. Mace*
11. Firearm*
12. Watchclock or bar-code reader (Detex equipment)
13. Camera
14. Whistle
15. Binoculars

Tools of the trade must be in working order before going on duty. When operating a patrol vehicle, it may be necessary to carry flares, first aid equipment, jumper cables, and a camera.

PERSONAL APPEARANCE

It is important to present a neat, clean and well groomed image. Uniforms should be pressed and tailored properly. Unless otherwise permitted, black, round-toe shoes should be worn. The uniform should be blue, black, or tan with a badge, patch, and name tag. For safety purposes and public relations, avoid wearing excessive jewelry (necklaces, ear rings, etc.). Excessive hair length and facial hair is discouraged. Again, the agency that employs a protection officer will specify grooming standards.

PATROL METHODS

There are two basic methods of patrol: foot and vehicle. Both are usually performed in uniform. Some assignments may require patrolling on bicycles or scooters. However, whether patrol is conducted by foot or vehicle, these are recommended procedures for effective patrol.

Foot Patrol

Foot patrol is a common method of detecting crime or safety hazards. While on foot patrol, the protection officer will be performing an invaluable public relations task. A friendly and understanding officer will engender confidence and respect on the part of the public. Foot patrol provides the protection officer with the opportunity to know the people and the business. The following suggestions are offered to assist foot patrol.

---

*Depending upon post assignment, training, and state laws.

1. *Avoid unnecessary conversations*. When walking a beat or area, do not engage in unnecessary or lengthy conversations with employees or patrons. Avoid being distracted from your duties. Be friendly and available without being overly solicitous; remember, a protection officer has a schedule to keep, especially if Detex rounds are required.

2. *Vary your patrol routine*. If patrolling a parking lot, housing complex, or other industrial or commercial setting, avoid walking the same pattern every day. Vary your route and avoid being predictable.

3. *Be observant*. While patrolling, look for hazards. Look behind and under things such as trash containers. Look for broken windows, leaky pipes, faulty wiring, or fencing. Check storage rooms, closets, stairwells, roofs, and other areas where safety or crime hazards may exist. Report suspicious vehicles or persons. Make note of license plate numbers and physical descriptions of vehicles which seem suspicious.

4. *Be cautious*. Approach each suspicious person or vehicle with caution. Do not get too close unless you feel it is safe. You can observe a great deal from a distance before approaching. At nighttime, approach each building with caution. An unauthorized open door or window may signal a burglary taking place. Stop momentarily, stand in a darkened area and listen for unusual sounds. If you suspect that a crime is taking place, do not enter alone. Call for assistance while keeping the building under observation. If a crime is occurring, do not make your presence known. Avoid grand entry statements such as, "Who's in here?"

5. *Take notes*. Record everything that may suggest a safety or crime problem (for example, open doors, unusual odors, broken windows).

6. *Communicate*. Carry a radio. Verify your location periodically with other officers or your base station; likewise, should you encounter a potentially dangerous situation, communicate the location and the type of situation (for example, suspicious persons).

Vehicle Patrol

Protection officers may perform their patrol duties in vehicles. Officers assigned to protect residential complexes, schools, or other large properties may be required to patrol by vehicle. A patrol vehicle should not be considered a vehicle of refuge, but rather a means of transportation and protection. Patrolling by vehicle may not be as thorough as foot patrol; therefore, it is suggested that the protection officer drive slowly and get out of the vehicle frequently in order to make spot checks. The following guidelines are suggested for vehicle patrol:

1. *Is the vehicle in good condition?* Be sure to check the vehicle for fuel, oil, and other fluids. Check the tire pressure and conditions. Carry and follow a checklist each time the vehicle is operated. The list should be posted on the dashboard to serve as a reminder. Report any vehicle deficiencies immediately.

2. *Open windows*. Open the windows of the vehicle whenever consistent with the weather. Turn your two-way radio down low so you can hear the sounds from the outside.

3. *Drive safely*. Operate the vehicle at normal speeds. In most cases, the protection officer will probably be driving on private property. There is no need to drive fast or pursue anyone. Obey all traffic regulations. Most patrol driving consists of stop-and-start, slow-speed driving with frequent backup. Many vehicular accidents occur during this type of driving.

4. *Keep awake.* Avoid fatigue especially if on evening patrol. Get out of the car; flex your muscles. Never use medication that may keep you awake or make you drowsy. Adequate rest is more important.

5. *Parking the vehicle.* When responding to an incident, park legally, unless an emergency exists. If you have emergency flashers, make sure that they are activated. Do not block driveways or pedestrian walkways. Do not park in areas reserved for the handicapped. If the vehicle is parked and must be left for a period of time, make sure it is locked. Never leave an unattended vehicle unlocked.

6. *Vary your patrol patterns.* As in foot patrol, vary your driving patterns. For example, instead of driving directly from parking lot A, to B, then to C, drive to lot B, to A and then to C, or from C, to A and then to B. In other words, change patrol routines, by implementing any number of different combinations that are available. Be unpredictable.

7. *Remember safety.* If you suspect that a crime is occurring at a particular location, do not park directly in front of it. Park a short distance away to allow you to observe and assess the situation. Do not park directly in front of a suspicious vehicle, park behind and at an angle to allow more visibility and personal safety. (See Figure 3–1.) As in foot patrol, do not enter a building or otherwise confront a hostile situation alone. Call for assistance or backup. Announce your location and situation to your dispatcher or fellow officer.

8. *Spotlights.* If your vehicle is equipped with a spotlight, it may be directed at unoccupied vehicles. Do not direct spotlights at moving vehicles; this may cause an accident.

9. *Transporting persons.* Unless otherwise directed by the agency, do not transport anyone in your vehicle while on duty.

10. *Arrests.* In cases where persons are arrested and transported, all prisoners will be frisked, hand-cuffed, and secured with a seat belt. Starting and Ending mileage should be reported to your dispatcher to protect against accusations of sexual misconduct.

FIGURE 3–1

Officer: *Unit 1 transporting one female arrestee to office. Mileage 24632 at 1615 hours.*

Control: *(should acknowledge and read back)*

Officer: *Unit 1 arrival at office mileage 24633 at 1617 hours.*

CROWD CONTROL

Crowds present a number of potential problems for the protectors of private property. Scheduled special events as well as spontaneous events have the potential for crowd formation. In either case, crowds can turn violent or become uncontrollable. Generally, crowds fall into the following categories:

1. Peaceful, planned crowds (concerts, fairs)
2. Peaceful, planned demonstrations (student demonstrations)
3. Labor strikes
4. Law-breaking or violent demonstrations
5. Rioting crowds (racial unrest)

Each one of these types of crowd presents a situation for which businesses and law enforcement must be prepared. Whether actively securing a crowd at a concert or passively protecting property from a passing crowd at a fair or demonstration, private protection officers must be prepared to deal with a variety of situations.

The training of protection officers for crowd control must come from the management of the premises protected or the employing protection company. The supervisor must adequately assess the protection of a given situation and make a recommendation to management. In turn, management chooses a course of action. Protection officers are then briefed regarding their responsibilities. Communication is a key component in preparing for a potentially unruly crowd.

When actively securing a crowd, the protection officer should know

1. The nature of the event.
2. Expected crowd behavior and the likelihood of possible opposition to the crowd (for example, proabortion vs. antiabortion).
3. Expected crowd size.
4. What profile he or she is expected to maintain (near the crowd, patrolling the crowd, fixed-posts).
5. What is being protected (people, property).
6. What areas are restricted and what areas are open.
7. Chain of command at the site.
8. What to do if the situation gets out of control (who to contact, how to adjust to the changing situation).
9. When to contact local law enforcement representatives.

To facilitate a group of protection officers controlling a crowd, a command post should be stationed on site and all officers must be able to get to it. The command post should include

1. A supervisor.
2 Radio communications.

3. An evacuation plan in case of emergency (including relevant maps and property configurations).
4. Emergency equipment.
5. Deployment plans.

Proper planning is also necessary to secure a passive site where crowds may be passing by private property and doing damage to it. Site managers should inform protection management regarding the specifics of the event. In addition, protection officers at the site need to know the following:

1. The boundaries and property configurations of the site they are protecting.
2. Who is allowed on the site and who is not (in some cases this may be very detailed).
3. Who to notify in case of an emergency.
4. What to do in the event of a disturbance.

At times crowds form spontaneously as the result of an unplanned event such as a fight, an accident, or a fire. These situations happen suddenly and can present a major challenge for a protection officer working alone at a site. The officer must quickly assess the situation and decide who to contact if the crowd is beyond his or her control. If help is needed, the officer will contact company management or the police. If the police are called, the officer's supervisor should also be notified.

The protection officer's primary job is to observe and report. Generally, observing and reporting can best be done by observing a crowd from a distance, or from a fixed post such as a raised platform in a parking lot. Systematic patrolling by foot or vehicle around the perimeter of an area and points of entry and exit is also recommended. Protection officials should be easily identified (distinctive uniforms) and always in communication with the command post or other officials. Officers should pay special attention to unruly groups, excessive profanity, the presence of alcohol in unauthorized areas and other signals of potential problems. A protection officer encountering a disturbance should not attempt to intervene alone, but should call for assistance. A show of force is preferable over lone intervention.

The procedure for a protection company to follow when trying to control a crowd is: planning, deployment, and communication. All personnel—site management, protection company management, and the protection officers—all need to work together in order to be prepared. Contingency planning is necessary so all parties can react quickly to a changing situation. And most importantly, constant communication is vital to the crowd-control operation.

WATCHCLOCK AND BAR-CODE SYSTEMS

A number of protection officers carry Detex equipment such as watchclocks and bar codes. A *watchclock* is a timepiece that contains a program tape or disc divided into time segments. The officer operates the watchclock by using keys located at specific vulnerable points (for example, storage rooms) throughout the site. When an officer makes a point visit during patrol, the key inserted into the clock records the visit. Watchclocks are used to verify that the officer visited the location.

Another, more state-of-the-art, system is referred to as *bar-code* technology. During patrol, the officer scans labels that look like black stripes. (See Figure 3–2.) These stripes represent numbers. As with keys in the watchclock system, the labels are located at critical points such as entrances and vaults. Protection

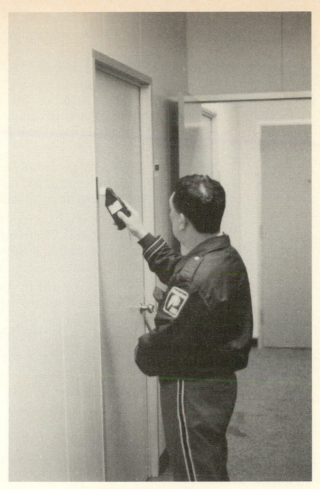

FIGURE 3–2

officers carry a hand-held scanner and a keypad for use when a bar-code label cannot be read.

While both systems are monitoring devices, their use also provides management with lower fire and burglary insurance premiums. Therefore, the use of these systems should be taken seriously by the protection officer. Before patrolling, the officer should check to make certain the equipment is working properly. Malfunctions should be reported immediately.

RESPONDING TO UNUSUAL INCIDENTS

Unusual incidents are those situations where a crime is being committed (or may be committed) or a major safety hazard exists. The following are examples of unusual incidents:

1. Unruly crowds such as found in labor disputes or concerts
2. Accidents involving vehicles or industrial equipment
3. Reports of crimes in progress
4. Chemical spills or other hazardous materials conditions
5. Natural disasters
6. Fires and explosions

Rather than attempt to define and address response procedures for every type of unusual incident, some general response guidelines follow.

1. *Identify the nature and extent of the problem.* Determine how much and what type of assistance is needed. If there are injuries, how severe are they? Do the police or other safety authorities need to be notified? At the time of your first assessment, it is better to overestimate than underestimate a problem; in other words, call for assistance and get help if in doubt.

2. *Communicate.* Notify your control center immediately. Keep in contact with your dispatcher. In some situations, a telephone may be used to afford more confidentiality in communications.

3. *Contain the area.* Once you have identified the problem, established communications, and summoned the required aid, keep unauthorized persons away from the scene. In crime scenes, it is important to protect evidence that may be used in court.

4. *Record the event.* Take notes of the incident. This is done once the incident has stabilized. In the case of an unruly crowd, attempt to determine who is inciting the crowd and why.

5. *Cooperate with public safety forces.* In the event that police or fire personnel are summoned, assist them in performing their duties. Offer information regarding movement in and around the property.

6. *The media.* The media may arrive at the scene of an unusual occurrence. News reporters may approach the protection officer and attempt to gather information. In such cases, it is best not to provide information that may be misleading or inaccurate; however, saying, "No comment," may suggest that the officer is hiding something. A good procedure is to inform the media that the incident is under investigation and as soon as it is complete, the media will be notified. The protection officer may also refer the media to management for additional information.

There may be an occasion where the protection officer will be confronted with an unusual incident such as a crime in progress; therefore, some further guidance is offered on what to do in these situations. What should be the response if someone is being assaulted? The protection officer may wish to react by taking immediate action through physical intervention. However, if the assailant is armed, or if the protection officer is outnumbered, physical intervention may be an unwise decision. If unarmed (most protection officers are not armed), the protection officer would be at a disadvantage. After all, if the protection officer becomes a victim, no one may be helped. At the other extreme, the protection officer who does nothing is neglecting his or her duty. The protection officer is not expected to sacrifice his or her life; however, an instant response is required. First, seek assistance. Communicate the location and type of situation. Second, keep the parties under observation from a safe distance. Third, attempt to communicate with the assailant. Warn that the police are on their way. Blowing a whistle, yelling "police" or other verbal warnings may be enough to persuade the offender to leave. Professional police hostage negotiators often attempt to de-escalate a situation by keeping the offender talking. Remember, most offenders are opportunistic and are not interested in being detected or caught; therefore, these stall tactics may be enough to force the offender to flee. There are no definite rules on what to do in these situations; however, be calm, observe, report, and communicate.

INVESTIGATION AND EVIDENCE

A protection officer may perform preliminary investigations on incidents ranging from thefts to injuries. Once dispatched to an incident, the protection officer should determine the nature of the problem, stand by to protect the scene, administer aid, comfort the victims, wait for the police or other specialists to arrive, and complete reports. The preliminary investigation also involves a careful search for evidence. During the initial investigation, consider the possibility that virtually everything found may be evidence. Evidence can be used to implicate someone in a crime, or can be used in later civil suits.

Although the protection officers are not police officers, they may encounter items or information that can be used by the police in conducting their investigations; therefore, it is important to understand the various types of evidence that can be found at a crime scene. Evidence can be provided by an eye witnesses (direct evidence) or circumstantial evidence such as tool marks or fingerprints. In most cases, the officer will look for circumstantial evidence.

Circumstantial Evidence

One valuable type of circumstantial evidence is fingerprints. There are generally three types of fingerprints found at the crime scene: latent prints, visible prints, and molded prints. *Molded prints* are usually found in soft substances such as grease, wax soap, and putty. *Visible prints* are those observed because the suspect may have touched a substance that serves as a printing medium (ink, blood, dirt). *Latent prints* are invisible impressions that must be developed by dusting with a special powder. Areas where prints may be found are windows, counter tops, desks, desk drawers and so forth. Thus, protecting the scene is important if the police are needed to take prints.

Another type of circumstantial evidence is tool marks. If you encounter a burglary where forced entry has been made, look for evidence of tool marks. Various types of restrictions and impressions are made by certain tools (such as hammers and screwdrivers). Take photographs and preserve the scene. Notify the police and your supervisor. It may be possible to match the marks with certain tools.

Other types of physical circumstantial evidence include shoe prints, tire prints, stains made by blood and other bodily fluids, fabrics, and virtually anything left at a crime scene. If responding to a crime or accident scene, make sure nothing is moved or touched. If you observe something that may be considered evidence, do your best to preserve it. Depending on your training, your role may include interviewing witnesses and victims. If you are expected to take reports, get as much information as possible. The physical characteristics forms (Figures 3–3 and 3–4) are presented to assist you in taking statements from crime victims or witnesses (Cox and Brown, 1992).

In cases of sexual assaults or violent crimes, it may be best not to confront the victim until the police or other specialists arrive. In such instances, assure the victim that help is coming. Attempt to make the victim comfortable and call for medical assistance. Do not try to question the victim since he or she may be in shock or otherwise under stress. If the victim wants to talk, you may take whatever notes or information is necessary.

In summary, the protection officer's responsibilities at the scene of a crime or other unusual incident is to

1. Assist the injured and request assistance.
2. Protect the scene.
3. Look for evidence.

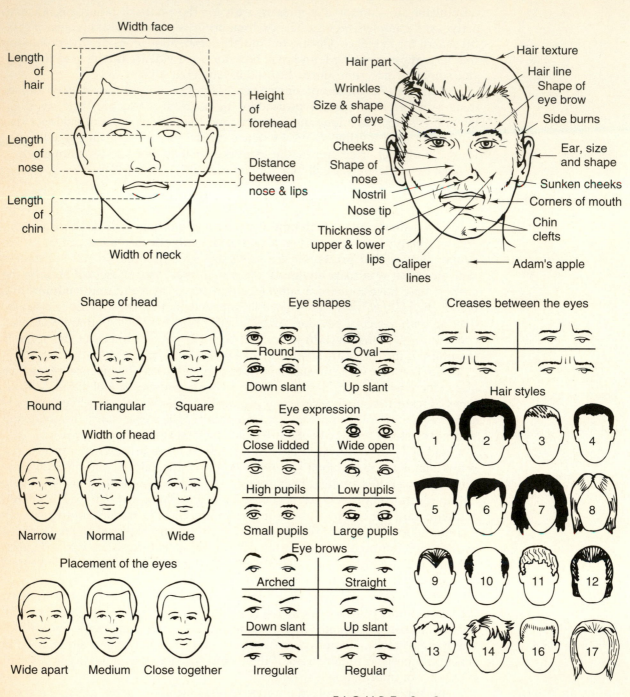

FIGURE 3-3

Physical Characteristics
Summary For Criminal Identification . . .

INSTRUCTIONS: Complete the following form with as much detail as possible. Include any memorable fact or item which caught your attention. Even the smallest detail could be important. Please print.

RACE:

☐ Caucasian   ☐ Hispanic   ☐ African American   ☐ Asian or Pacific   ☐ American Indian
                                               (non-Hispanic)   Islander   or Alaskan Native

                Other _____

SPEECH ACCENT (Nationality): _____

HEIGHT: _____ ft. _____ in.   WEIGHT: _____ lbs.

BODY BUILD: ☐ Fat   ☐ Average   ☐ Slim   ☐ Stocky

EYES: Color _____   Glasses _____

HAIR: Color _____   Style _____   Beard _____   Moustache _____

SCARS: _____   Where _____

UNUSUAL CHARACTERISTICS: _____

_____

_____

_____

DISGUISE: _____

_____

FIGURE 3–4

4. Preserve and protect items that may be evidence.
5. Take notes and/or photographs of the scene.
6. Refrain from giving any unauthorized statements or opinions about the incident.

TESTIMONY

From time to time protection officers may be required to provide testimony. If the protection officer took an incident report, testimony may be required to verify the report, or any knowledge the officer may have concerning the event. The protection officer may be asked to testify in a criminal or civil case. There are generally two types of testimony: out-of-court and in-court testimony.

Out-of-Court Testimony

An example of an out-of-court testimony is a deposition. A deposition is a question and answer session between a witness and the attorneys. Attorneys from both sides are present and may ask questions of the protection officer. Depositions are out-of-court, sworn testimony which are officially recorded. Depositions are taken before trial as part of the so-called pretrial discovery process. Depositions can last an hour or several hours. They can be difficult because there is no judge to referee the process. If you, as an officer, are called for a deposition, meet with an attorney beforehand to discuss the expected testimony. Review all notes and reports, but do not give opinions or lengthy explanations. Do not take any materials to the deposition unless requested. What is said in a deposition can be a factor in determining whether the case goes to trial or is settled. Therefore, listen to each question and answer. If you don't know the answer, say, "I don't know." Don't try to give answers that may please the attorneys. If tired or otherwise in need of a break to speak to your attorney, say so! Don't be shy or hesitant. Attempt to answer every question to the best of your ability. Do not volunteer information not requested.

Example:

Attorney:	*Did you take the report regarding the assault on Mrs. Jones?*
Protection Officer:	
Preferred response	*Yes, I did.*
Unacceptable response	*Yes, I did and I have taken a lot of other reports that have occurred in the parking lot…*

In the second response, the protection officer is volunteering information not asked. This may open up more questions that the protection officer will be expected to answer. Let the attorney ask the questions first. Don't try to help the attorney! If, during the testimony, you would like to add something or make a correction, make it known. Something can be added anytime by simply saying, "There is something I forgot to mention," or, "I would like to add to my last answer."

After the deposition has been completed, there will be an opportunity to read and sign it. Corrections can be made if errors are found; however, be advised that any corrections may be commented on by opposing attorneys at the time of trial. It is preferred that the best testimony be provided at the time of the deposition.

In-Court Testimony

In-court testimony refers to the trial process. The trial is often referred to as "show time." During the trial, a protection officer may be testifying before a jury and/or a judge. There may be other spectators in the courtroom as well. Dress and personal appearance are important during the trial process. The behavior, tone of voice, and posture of the person on the stand are all evaluated by the jury or judge.

If called to testify in a case, the protection officer will be sworn in and most likely testify as a factual witness. That means, the protection officer will testify about what is known or observed. The protection officer will not give opinions since opinion testimony is reserved for expert witnesses. If a deposition was given at an earlier date, the officer must be prepared to defend it. Questions may be asked according to testimony given at the deposition; therefore, the officer should review the deposition (as well as any other notes or reports that have been prepared) prior to trial.

Should you, as a protection officer, be required to give trial testimony, attempts will be made by the opposing attorney to confuse you or provoke anger. Do not argue with the attorney. Be polite and listen carefully to each question. Remember, the judge is there to see that you're treated fairly. If a question is asked and an objection is made by your attorney, remain quiet until the judge rules on the objection. If the objection is sustained by the judge, it means that the attorney must ask a different question and the original question does not require an answer. If the objection is overruled, the protection officer will be instructed to answer the original question. If there is no objection, then the protection officer is expected to answer.

Example:

Question by opposing attorney:	*Officer Jones, isn't it true that you warned management about the security at the hotel?* *(short pause and wait for an objection)*
Objection by your attorney:	*Your Honor, I object to this question as being irrelevant, vague, and unfounded.*
Judge:	*Overruled (Officer Jones must answer)* *or*
Judge:	*Sustained (Officer Jones need not answer). The opposing attorney must ask a different question.*

After providing testimony, you will be excused by the judge. Proceed directly from the courtroom. During court recess, do not engage in conversations with the jury or other spectators. Keep to yourself. Your attorney may wish to speak with you, but that should be the limit of your conversations.

In summary, always be prepared before testifying. Make sure you have had enough sleep and are not taking any medication which may affect your reasoning. Dress appropriately (business suit or uniform) and be well groomed. Remember, the image you project may have an effect on how the judge or jury perceives you as a person and as a professional.

RADIO COMMUNICATIONS

Radio communications are used by hundreds of different types of agencies. The two-way radio is one of the protection officer's most important pieces of equipment. Without a properly working radio, the protection officer's ability to report incidents as well as her or his personal safety is jeopardized. Radio transmissions should be brief and clear. The Federal Communications Commission (FCC) is charged with responsibility for the legal and efficient use of radio frequencies. Profanity and other types of unwarranted language are prohibited.

There are some recommended techniques to be followed when using a radio. First, prior to broadcasting, make sure the radio is working properly by verifying with the base station or fellow officers. Simply give your name or identification number and the announcement "radio check." Wait for a response. Second, when speaking into the microphone speak slowly and clearly approximately 2–3 inches from the microphone. Third, when using the radio, do not start talking until initial contact with the receiver has been established.

Example:

Protection Officer:	Unit Adam to control.
Control:	Unit Adam go ahead.
Protection Officer:	There is a broken window at...

Do not start talking with the hope that someone is listening (with the possible exception of an emergency). Also, if you're broadcasting long messages, pause periodically to make sure the message is being received. Fourth, before broadcasting, listen to the radio, and make sure no one else is broadcasting. Fifth, it is recommended that radio codes be used. The use of codes reduces broadcast time making the communication more simple. Codes also provide more secrecy in communications. A protection officer's agency may develop its own codes, use other systems (see Figure 3–5), or select parts of one system. Obviously, some of the codes are associated more with law enforcement. The protection officer's employer will most likely decide what codes, if any, to use. Whatever system is used, make sure you know them well! Sixth, do not use radios for personal conversations. Likewise, do not identify others by name if at all possible. Use unit numbers or other codes representing posts or patrol assignments.

Example:

Base station:	Bravo post 10-21 Base
Bravo post:	10-4

Finally, always keep a portable two-way radio on hand. If there is a need to broadcast, make sure that transmissions are possible in that location. Avoid transmitting near operating machinery or in areas associated with radio interference. The protection officer should know the areas that cause transmission interference.

The radio is often your only contact with headquarters As a protection officer you will rely heavily on the dispatcher to provide assistance and protection when needed. Radio broadcasting must be clear and understandable. Keep calm in order to avoid transmitting undue anxiety. Impersonal transmissions are standard operating procedure. Limit transmissions to only that which is essential and directly related to the official business at hand.

The following guidelines should be followed for effective broadcasting results when using the police radio (Adams, 1990).

1. Practice courtesy. It is contagious and will make work much easier for those who use the radio as well as those who must listen to it during their entire tour of duty.
2. Broadcast station-to-station only. Mike-to-Joe transmissions are not allowed. Such communication should be held in the context of personal meetings or telephone calls.
3. Humor and horseplay have no place on the frequency. Rude sounds and sarcastic comments emanating from anonymous users of the radio frequency may sound funny for the moment, but they are immature, unlawful, and they may someday interfere with a transmission that could mean the difference between life and death. Leave the humor to those employed specifically for the purpose of entertainment.
4. Avoid any use of the radio for personal conflicts such as "chewing out," arguments, or sarcasm.
5. Keep all transmissions brief and to the point. Use the telephone for lengthy messages.

10-1	Receiving poorly	10-23	Stand by
10-2	Receiving well	10-25	Do you have contact with _____?
10-3	Stop transmitting	10-27	Check _____
10-4	OK, or acknowledge	10-28	Registration request
10-5	Relay	10-29	Check for stolen or wanted
10-6	Busy	10-30	No records or wants your subject
10-7	Out of service	10-31	Subject has record, but not wants
10-7B	Out of service at home	10-32	Subject wanted. Are you clear to copy?
10-8	In service	10-33	Standby. Emergency traffic only.
10-9	Repeat	10-34	Resume normal radio traffic
10-10	Out of service, subject to call	10-35	Confidential information
10-11	Transmitting too rapidly	10-36	Correct time
10-12	Officials or visitors present	10-37	Name of operator on duty
10-13	Weather and road conditions	10-39	Message delivered
10-14	Escort or convoy	10-40	Is _____ available for telephone call?
10-15	En route with prisoner	10-40A	Is _____ available for radio call?
10-16	Pick up prisoner	10-42	Pick up officer
10-17	Pick up papers	10-45	Service your equipment
10-18	Complete present assignment as quickly as possible	10-46	Standby. I am moving to a better location.
10-19	Return or returning to the station	10-48	I am now ready to receive information.
10-20	Location, or what is your location?	10-49	Proceed to _____
10-21	Call your station or dispatcher by telephone	10-86	Traffic check (Do you have any messages for this unit?)
10-21A	Advise my home I will return at _____	10-87	Meet _____ at _____
10-21B	Call your home	10-88	What phone number should we call for station-to-station call?
10-21T	Reply via teleprinter	10-97	Arrived at the scene
10-22	Cancel last message or assignment	10-98	Completed last assignment

F I G U R E   3 – 5   Radio codes for emergency services.

6. Profane and indecent language is unlawful.
7. Transmit only essential messages. Avoid asking for the time of day just to hear the radio. Secure routine information such as time to eat lunch and other scheduled matters, either before going out in the field or by telephone.
8. Be completely familiar with your department's radio procedures and the individual items of equipment.
9. Assume a personal responsibility for correct and intelligent use of the radio by your department.
10. Use radio code for brevity and accuracy when practicable.

Conclusion
The patrol function is often considered the backbone of any protection function. An alert well trained protection force can make the difference between a safe and unsafe environment. Protection officers must also be service oriented. In other words, business invitees not only expect competence, but courtesy. A professional attitude and appearance on the part of the protection officer will promote consumer confidence in the business.

Discussion Questions
1. Explain why it is important to vary a foot or vehicle patrol pattern.
2. What is the recommended security response if a burglary in progress is suspected?
3. Discuss the recommended steps in conducting a preliminary investigation of a crime.
4. Identify the differences between in-court and out-of-court testimony.
5. What are the advantages of using codes during radio transmissions?
6. Explain the purpose of a watchclock during patrol .
7. Discuss the three types of fingerprints that can be found at a crime scene.
8. While on foot patrol, a protection officer observes a burglary taking place in a remote section of a parking lot. The victim is also being beaten by several suspects. The suspects are armed with handguns (the officer is not armed). Based on these facts, what would be the preferred protection-officer response?

References
ADAMS, THOMAS F., *Police Field Operations*, (2nd ed.), pp. 127–9. Englewood Cliffs: Prentice-Hall Inc., NJ, 1990.
COX, CLERIC R., AND JERROLD G. BROWN, *Report Writing For Criminal Justice Professionals*. Cincinnati, Ohio: Anderson Publishing, 1992.

Chapter 4

Report Writing

After studying this chapter you should be able to explain the following:

Basic rules of grammar

Spelling techniques

Commonly misspelled words

Appropriate use of abbreviations

Listening skills

Note-taking techniques

Daily activity reports

Incident reports

Miscellaneous reports

INTRODUCTION

Preparing a written report is one of the most important responsibilities of a protection officer. For example, a well-written report presents the facts in order by stating what happened, who was involved, and when it happened. Since written reports help establish facts they can be presented in court as evidence. Written reports are invaluable in documenting the who, what, when, where, and how of any circumstance. Good report writing takes practice and patience. There are a number of skills required in writing effective reports. The protection officer must be organized and have a basic understanding of proper grammar and spelling. This chapter offers guidelines in preparing written reports.

BASIC GRAMMAR AND SPELLING

Correct grammar and spelling is important in report writing. Incomplete sentences or poor spelling can affect the quality of a report. Your writing style is a reflection of your competence as a protection officer. Always have a dictionary nearby when preparing reports.

Grammar

The proper use of grammar requires knowledge of the basic parts of speech: noun, pronoun, verb, adverb, adjective, preposition, and conjunction. The following is a description of each:

Noun is the name of a person, place, thing, or quality.
Examples: officer, England, book, loyalty

Pronoun stands in place of a noun.
Examples: he, she, it, they, them, their, you, your

Verb expresses action or state of being.
Examples: ride, run, hide, be, seem

Adverb modifies or limits a verb, adjective, or another adverb. It is sometimes an intensifier.
Examples: carefully, quickly, very, too

Adjective modifies or limits a noun or pronoun.
Examples: red, curly, dark, smart, quick

Preposition shows the relationship between a noun or pronoun and some other word in the sentence.
Examples: to, in, on, between, among, with, of

Conjunction joins or connects words or groups of words.
Examples: and, but, or

In writing reports, you will be writing about past events; therefore, you will often use the past tense to describe previous incidents. For example, if an employee reports that a slippery floor caused her to fall, you would write: Employee stated she slipped on the floor. Slipped is the past tense of slip (a verb). You would not write that the employee slip on the floor or is slipping on the floor. Slipping suggests the present act of falling.

Sentences

A *sentence* is a group of words which form a complete thought and contain a subject and a predicate. A *complete subject* is a word or group of words that tells what the writer is saying. A *complete predicate* is a part of the sentence that tells us about the subject. In other words, it tells what the subject is doing or what is happening to the subject.

Complete Subject	Complete Predicate
the victim	was assaulted
she	works at the hotel
the officer	was injured
glass and debris	covered the parking lot

A *simple subject* is the particular word (a *compound subject* is the word or a group of words) in a complete subject about which something is said. A *simple predicate* is the main word (or words if a *compound predicate*) in the complete predicate. Nouns are simple subjects and verbs are simple predicates.

In writing a sentence, concentrate on who did what to whom, as well as when, where, and how. Unless asked to give an opinion, there is no concern as to why an event occurred. Remember that reports are based upon factual events, rather than opinions.

In the following sentence the subject, verb, and object are:

The	*supervisor*	*gave*	*an*	*order.*
	who	did		what
	Subject	Verb		Object

The supervisor (who gave the order) is the subject of the sentence. The verb *gave* tells us what he or she did, and the direct object *order* describes what was done. Occasionally, you may write a report using an indirect object. The indirect object of a sentence refers to any noun or noun substitute that states to whom or for whom (or to what or for what) something is done.

Examples:

She gave the officer an order.

He gave an order to the officer.

In the second example, note that *to the officer* is a prepositional phrase. A prepositional phrase begins with a preposition such as *to, by, of, for, in, on*, or *between*. A prepositional phrase is not a sentence and therefore cannot stand alone. It is a group of related words without both a subject and a verb.

Many grammatical mistakes made in report writing result from the misuse of the verb. Some verbs do not need objects.

	Verb		*Noun*
He	appeared	injured.	
	Verb	Adjective	

Verbs can be active or passive. An active verb shows the action that the subject took.

Burglars	often	steal	jewelry.

In its passive form, the subject is acted upon.

Jewelry	is	often	stolen	by	burglars.
The	report	was	written	by	Officer Jones.

Many mistakes involving verbs result from misuse of the passive state. To avoid this problem, consider the following suggestions:

1. If you are preparing a report, refer to yourself as I, *the writer,* or *Officer Jones.*
   Example: I received a complaint about an unruly person in the lobby. When I arrived, I was told that the person had left.
2. You can also write short sentences.
   Example: Received a complaint about an unruly person in the lobby. Arrived at the lobby. The person left. Checked the area. Person not found.

Independent and Dependent Clauses

In writing sentences, it is important to know the difference between an independent and dependent clause. An *independent clause* contains a subject and predicate. It is able to stand alone.

She   ran.
He   fell.

A *dependent clause* contains a subject and predicate but is not able to stand alone. A dependent clause is used as an adverb, an adjective, or sometimes as a noun. A dependent clause is not a sentence. A dependent clause is usually introduced by such words as *when, if, because, since, that, which, who, so, although, after, unless, until, before, whom, why, whether,* and *while.*

Example:
Before   the   officers   arrived
Because   he   was   injured

In the preceding example, the clause cannot stand alone. Both are incomplete sentences. To complete the sentence, you would need to clarify what occurred before the officers arrived, and what took place regarding the injury.

Example:
   The suspect fled   before the officers arrived.
Because he was injured,   the victim was taken to the hospital.

Punctuation Problems

In report writing, it is the practice of some writers to use too many commas. A common rule is to use commas between independent clauses when they are joined by *and, or, nor, but, for* or *yet.*

Example:
Two suspects were hospitalized, but no officers were injured.

Note that, *Two suspects were hospitalized,* is a complete sentence. *No officers were injured* is also a sentence. Both are joined by *but.*
   When a long dependent clause precedes an independent clause, a comma follows the dependent clause.

Example:
When the suspects ran from the parking garage, the officer radioed for help.

When the independent clause comes first, usually no comma is necessary.

Example:
She requested a doctor when she realized the seriousness of the injury.

In the second example, the words tend to flow without a pause. In the first example, there is a pause between *garage* and *the*.

There are other uses of the comma. In addresses, commas are used to set off towns, counties, and states.

Example:
She lived at 110 2nd Street, Chicago, Illinois 55555

Note there is no comma used before the ZIP code. In dates, commas are needed to separate the day, month, and year.

Example:
The shooting occurred on Tuesday, April 7, 1992.

Some alternate forms are used. Military and European usage sometimes reverses the order, omitting the commas.

Example:
7 April 1992

Commas may separate the words in a series before the final conjunction *and*, or *nor*.

Example:
The room was dark, cold, and damp.

A comma can be used before the conjunction *and*, but it is not necessary. Be consistent in using commas throughout the report. Commas set off direct quotations.

Example:
The suspect said, "I didn't do it."
The officer told his supervisor, "I quit."

Commas are used to set off titles or degrees.

Example:
John Smith, Ph. D.

The semicolon is rarely used in private-protection report writing. When it is used, the *semicolon* separates long or confusing items in a series. The semicolon separates independent clauses when no coordinating conjunction is used. Transition words such as *however, hence, therefore, moreover, consequently*, and *in fact* require a semicolon preceding and a comma after the word when they are used to connect two independent clauses.

Example:
The supervisor had previously warned the officer about being late; however, the officer failed to listen.

Colon.   The *colon* usually means that something will follow.

Example:

Protection officers are required to carry the following equipment: flashlight, radio, notebook, handcuffs, baton, and firearm.

In writing memos, the colon is used after key words.

Example:

    To: Chief Jones
  From: Officer Smith
Subject: Training

Apostrophe.   The *apostrophe* is used to show possession. It can also show that something has been omitted. There are a number of uses for an apostrophe. An apostrophe is used followed by an *s* to show possession when the noun does not end with an *s* sound. If the noun has an *s* at the end, add the apostrophe.

Examples:

The suspect's shirt
The officer's schedule
Someone's job
Another's flashlight

The apostrophe can show omission of one or more letters.

Example:

I have received two weeks vacation.
I've received two weeks vacation.
He can not go.
He can't go.
He is not working.
He isn't working.

Parentheses.   *Parentheses* are used to separate certain word groupings. In writing a sentence there may be a group of words that do not flow with the sentence, but are necessary to add meaning to the sentence.

Example:

The officer (who has received many honors) was awarded a distinguished service medal.

Quotation Marks.   There are several uses of *quotation marks.* The most common is to set off a direct quotation. The quotation marks are placed before the first word of the quotation and after the final punctuation.

Examples:

The suspect stated, "I shot him."
The victim yelled, "Don't hurt me."

Sometimes quotation marks are used to indicate an unusual or special statement.

Example:

The captain said that all officers completing the training program would be given a "raise."

Spelling

Proper spelling is an important part of report writing. Misspelled words can affect your credibility as a protection officer. Some common errors in spelling are:

1. Spelling the names of victims, suspects, or witnesses incorrectly.
2. Misspelling words commonly used in the protection industry.
3. Making gross spelling errors in everyday words.
4. Using words that sound the same but mean something entirely different.

Many spelling errors can be avoided by carefully reviewing your report before it is submitted. Reviewing your report requires that you proofread your writing to make sure that the right words are used. Verify the spelling of all names used in the report. Invest in a pocket dictionary.

Spelling errors also result from poor vocabulary or pronunciation skills. It is important to have a standard pocket dictionary available when preparing reports. Another way to avoid spelling problems is to keep a list of commonly misspelled words. The following list of words are commonly spelled incorrectly in police and security reports. Review the list and know the meaning of each word.

Spelling errors can be reduced by following basic rules. Remember that words are made up of vowels and consonants. Vowels are *a, e, i, o, u*. Consonants are all other letters. In deciding how to add *ing* to a word, you must know about vowels and consonants. Vowels can be either long or short. Long vowels are pronounced the same as the letter. In long vowels, drop the *e* and add *ing*, do not double the consonant.

Example:

give   giving   not giveing
rape   raping   not rapping or rapeing

In the preceding example, *e* is. To add *ing* drop the *e*.

There are a number of words which have tricky letter combinations. There are some rules to remember to assist in spelling properly.

Rule: *Use* i *before* e *except after* c *or when pronounced* ay *as in neighbor.*

This rule is true most of the time.

Examples:

believe, chief, yield, receive, reign, receipt
Some words violate this rule.

Examples:

weight, height, foreign

Forming plurals causes some confusion in report writing. The following rules are suggested to reduce this confusion:

Rule: *Most plural forms are made by adding* s. *Example:*

Singular	Plural
Auto	Autos
Gun	Guns
Book	Books

Rule: *Nouns ending in* s, sh, ch, x *or* z *form the plural by adding* es. *Examples:*

Singular	Plural
Bush	Bushes
Church	Churches

Rule: *Nouns ending in* o *preceded by a vowel add* s. *Example:*

Singular	Plural
Piano	Pianos
Rodeo	Rodeos

Rule: *Some nouns ending in* o *preceded by a consonant add* s; *others add* es. *Example:*

Singular	Plural
Cargo	Cargos or cargoes
Hero	Heroes

Rule: *Nouns ending in* y *preceded by a consonant, change the* y *to* i *and add* es. *Example:*

Singular	Plural
Factory	Factories
Duty	Duties
County	Counties

Rule: *Nouns ending in* y *preceded by a vowel usually add* s. *Example:*

Singular	Plural
Day	Days
Alley	Alleys

Rule: *Some nouns ending in* f *or* fe *change the* f *or* fe *to* v *and add* es. *Example:*

Singular	Plural
Life	Lives
Thief	Thieves
Leaf	Leaves

Rule: *Some nouns form a plural by changing the vowel. Example:*

Singular	Plural
Man	Men
Woman	Women

Rule: *Some nouns are spelled the same in both the singular and plural.*

Incorrect use of the English language in conversation frequently leads to spelling errors. Some common problems fall into the following categories:

1. Incorrect pronunciation.
2. Misconstrued meanings of words.

3. A lack of understanding about homonyms or words which sound alike but mean something different and are spelled differently (for example, *bear* and *bare*).
4. Slang or jargon.

Verbs.   Learn to distinguish *regular verbs* (verbs that form the past tense and the past participle by adding *ed* or *d* to the form of the present tense) and *irregular verbs* (verbs that do not form the past tense and the past participle in the regular way). Examples of regular verbs:

Present Tense	**Past Participle**
(present time)	(with *have, has, had*)
call	called
join	joined
build	built

Examples of irregular verbs:

drive	driven
go	gone
burst	burst

Past Tense
(past time)
called
joined
built

drove
went
burst

Verbs generally are not a serious problem in writing. However, irregular verbs are responsible for most verb errors. A list of troublesome verbs are provided (Cox and Brown,1992).

Present Tense	**Past Tense**	**Past Participle**
(*present time*)	(*past time*)	(with *have, has, had*)
be	was	been
beat	beat	beaten
begin	began	begun
bid (offer to buy)	bid	bid
bid (command)	bade	bidden, bid
blow	blew	blown
catch	caught	caught
climb	climbed	climbed
cut	cut	cut
dive	dived	dived
do	did	done
draw	drew	drawn
drive	drove	driven
eat	ate	eaten
fall	fell	fallen
fly	flew	flown

Present Tense (present time)	Past Tense (past time)	Past Participle (with have, has, had)
forget	forgot	forgotten
get	got	got, gotten
give	gave	given
go	went	gone
hang (a picture)	hung	hung
hang (a criminal)	hanged	hanged
lay (to place, to put)	laid	laid
lend	lent	lent
let	let	let
lie (to recline)	lay (not laid)	lain (not laid)
lie (tell a falsehood)	lied	lied
prove	proved	proved
ride	rode	ridden
ring	rang	rung
rise	rose	risen
say	said	said
see	saw	seen
set	set	set
shake	shook	shaken
shine (give light)	shone	shone
shine (polish)	shined	shined
tell	told	told
throw	threw	thrown
understand	understood	understood
wear	wore	worn
weep	wept	wept

Examples:

Present Tense: **Drive** the car.

Past Tense: I **drove** to the scene.

Past Participle: I **was driven** to the scene.

Slang and Jargon. When writing effective sentences, try to keep them under 15 words long. Sentence length can be decreased by eliminating unnecessary words. Avoid using slang or jargon because these are another area of possible confusion in writing. A good rule is: Use slang only in the case of direct quotes, and do not attempt it unless it is simple and easily understood. Jargon is the use of a specialized language of a profession.

Proofreading. To write and prepare effective reports it is important that you develop good proofreading skills. Be especially alert when proofreading your own work. Proofread twice—once for content and the other for grammar and spelling. Keep your own list of commonly misspelled words and misused vocabulary as a reference when proofreading.

The following is a list of commonly misspelled words (Cox and Brown, 1992;)

abduction	achievement	altercation
accelerated	acquire	among
accessories	acquitted	analyze
accident	affidavit	apparatus
accommodate	all right	apparent

arguing
argument
arson
assault
belief
believe
believes
beneficial
benefitted
bureau
burglary
category
coercion
coming
commission
comparative
complainant
conscious
conspiracy
controversial
controversy
conviction
corpse
counterfeit
criminal
defendant
define
definitely
definition
describe
description
disastrous
dispatched
disposition
drunkenness
effect
embarrass
embezzlement
emergency
environment
exaggerate
existence
existent
experience
explanation
evidence
extortion
fascinate
forcible
fraudulent
height
homicide
interest

intimidation
intoxication
its (it's)
investigation
juvenile
larceny
led
legal
lieutenant
lose
losing
marriage
marshall
mere
necessary
occasion
occurred
occurrence
occurring
offense
official
opinion
opportunity
paid
particular
patrolling
pedestrian
performance
personal
personnel
possession
possible
practical
precede
precinct
premises
prejudice
prepare
prevalent
principal
principle
privilege
probably
procedure
proceed
professor
profession
prominent
prosecute
prostitution
pursue
pursuit
quiet

receive
receiving
recommend
referring
repetition
resistance
rhythm
robbery
sabotage
scene
seize
sense
sentence
separate
separation
sergeant
serious
sheriff
shining
similar
statute
strangulation
studying
subpoena
succeed
succession
suicide
summons
surprise
surrender
surveillance
suspect
suspicion
techniques
than
then
their
there
they're
testimony
thieves
thorough
to / too / two
traffic
transferred
trespassing
truancy
unnecessary
victim
warrant
write

ABBREVIATIONS

Your reports will often include the use of abbreviations. This may vary widely from agency to agency. Generally, use the locally accepted form and be consistent. Abbreviations will save you time but personal or unknown abbreviations should not be used in a professional report. Use the following rules for abbreviations:

1. Spell out all titles except Mr., Mrs., Ms., Miss, Dr., and St. (*saint*, not *street*).
2. Spell out *Street, Road, Park, Company*, and similar words used as part of a proper name or title.
3. Spell out personal names (*William*, not *Wm.*).

The following lists are for reference only and can be helpful in note taking.

Suspect Abbreviations

List in order of race-gender-age. For example, WFA (White-Female-Adult); BMA (Black-Male-Adult).

Standard Abbreviations

Dates

Jan.	Apr.	Jul.	Oct.
Feb.	May	Aug.	Nov.
Mar.	Jun.	Sept.	Dec.

Mon.	Thu.	Sun.
Tue.	Fri.	
Wed.	Sat.	

1st	3rd	5th	7th	9th
2nd	4th	6th	8th	10th

Time

Year	Yr.	Week	Wk.	Minute	Min.
Years	Yrs.	Weeks	Wks.	Minutes	Mins.
Month	Mo.	Hour	Hr.	Second	Sec.
Months	Mos.	Hours	Hrs.	Seconds	Secs.

Measurements

Inch	In.	Mile	Mi.	Pound	Lb.
Feet/foot	Ft.	Gram	Gr.	Pounds	Lbs.
Yard	Yd.	Kilogram	Kg.	Ounce	Oz.
Kilometer	Km.	Dozen	Doz.	Measurement	Meas.
Length	L.	Height	Hgt.	Width	W.
				weight	Wt.

Other Common Abbreviations

Administration	Admin.	Arrest	Arr.
Against	v. or vs.	Assistant	Asst.
All Points Bulletin	APB	Assist Outside Agency	AOA
Also Known As	AKA	Attorney	Atty.
Amount	Amt.	Attempt	Att.
Approximate	Approx.	Attempt to Locate	ATL

Birthplace	BPL	Left Front	LF
Building	Bldg.	Left Hand	LH
Burglary	Burg.	Left Rear	LR
California Driver's Lic	CDL	License	Lic.
Captain	Capt.	Lieutenant	Lt.
Caucasian	Cauc.	Manager	Mgr.
Central	Cen.	Manufacturing	Mfg.
Channel	Chan.	Maximum	Max.
Company	Co.	Mechanical	Mech.
Construction	Constr.	Medium	Med.
Convertible	Cvt.	Memorandum	Memo
Court	Ct.	Middle Initial	MI
Crime Scene		Miles per hour	mph
Investigator	CSI	Military Police	MP
Date of Birth	DOB	Misdemeanor	Misd.
Dead on Arrival	DOA	Modus Operandi	MO
Defendant	Def.	National Crime	
Department	Dept.	Information Center	NCIC
Department of Motor		No Further Description	NFD
Vehicles	DMV	No Middle Name	NMN
Describe	Desc.	Not Applicable	NA
Description	Descp.	Northbound	N/B
Detective	Det.	Number	No.
Director	Dir.	Officer / official	Off.
District	Dist.	Opposite	Opp.
Division	Div.	Organization	Org.
Doctor	Dr.	Package	Pkg.
Doctor of Veterinary		Page	p.
Medicine	D.V.M.	Pages	pp.
Doing Business As	DBA	Passenger	Pass.
Drivers License	DL	Permanent	
Driving Under the		Identification Number	PIN
Influence	DUI	Pieces	Pcs.
Driving While		Pint	Pt.
Intoxicated	DWI	Place	Pl.
Eastbound	E/B	Place of Entry	POE
Enclosure	Encl.	Point of Impact	POI
Engineers	Engrs.	Police Officer/	
Example	Ex.	Probation Officer	PO
Executive	Exec.	Post Office	P.O.
Federal	Fed.	Quantity	Qty.
Freight	Frt.	Quart	Qt.
Gauge	Ga.	Received	Recd.
General Broadcast	GB	Required	Req.
Government	Govt.	Right	R
Headquarters	Hdq.	Right Front	RF
High Voltage	HV	Road	Rd.
Highway	Hwy.	Right Rear/Rural Route/	
Hospital	Hosp.	Railroad	RR
Identification	ID	School	Sch.
Junior	Jr.	Section	Sect.
Juvenile	Juv.	Secure	Sec.
Last Known Address	LKA	Sergeant	Sgt.
Left	L	Serial	Ser.

Shore Police	SP	Unknown	Unk.
Southbound	S/B	Vehicle Identification	
Subject	Subj.	Number	VIN
Superintendent	Supt.	versus	v. or vs.
Surface	Sur.	Veteran	Vet.
Symbol	Sym.	Volume	Vol.
Tablespoon	Tbsp.	Weapon	Wpn.
Technical	Tech.	Westbound	W/B
Teletype	TT	Wholesale	Whsle.
United States of America	U.S.A.		

State Abbreviations

Alabama	AL	Montana	MT
Alaska	AK	Nebraska	NE
Arizona	AZ	Nevada	NV
Arkansas	AR	New Hampshire	NH
California	CA	New Jersey	NJ
Colorado	CO	New Mexico	NM
Connecticut	CT	New York	NY
Delaware	DE	North Carolina	NC
District of Columbia	DC	North Dakota	ND
Florida	FL	Ohio	OH
Georgia	GA	Oklahoma	OK
Hawaii	HI	Oregon	OR
Idaho	ID	Pennsylvania	PA
Illinois	IL	Puerto Rico	PR
Indiana	IN	Rhode Island	RI
Iowa	IA	South Carolina	SC
Kansas	KS	South Dakota	SD
Kentucky	KY	Tennessee	TN
Louisiana	LA	Texas	TX
Maine	ME	Utah	UT
Maryland	MD	Vermont	VT
Massachusetts	MS	Virginia	VA
Michigan	MI	Washington	WA
Minnesota	MN	West Virginia	WV
Mississippi	MS	Wisconsin	WI
Missouri	MO	Wyoming	WY

Military Time.   Reporting times is much easier if you use the military version. This removes any confusion as to whether it is day or evening.

Regular	Military	Regular	Military
12:10 A.M.	0010 Hours	1:00 P.M.	1300 Hours
1:00 A.M.	0100 Hours	2:00 P.M.	1400 Hours
2:00 A.M.	0200 Hours	3:00 P.M.	1500 Hours
3:00 A.M.	0300 Hours	4:00 P.M.	1600 Hours
4:00 A.M.	0400 Hours	5:00 P.M.	1700 Hours
5:00 A.M.	0500 Hours	6:00 P.M.	1800 Hours
6:00 A.M.	0600 Hours	7:00 P.M.	1900 Hours
7:00 A.M.	0700 Hours	8:00 P.M.	2000 Hours
8:00 A.M.	0800 Hours	9:00 P.M.	2100 Hours

Regular	Military	Regular	Military
9:00 A.M.	0900 Hours	10:00 P.M.	2200 Hours
10:00 A.M.	1000 Hours	11:00 P.M.	2300 Hours
11:00 A.M.	1100 Hours	12:00 P.M.	2400 Hours
12:00 A.M.	1200 Hours		

Obviously, you should report events as precisely as possible. For example: 1:15 P.M. = 1315; 6:30 A.M. = 0630; 12:15 A.M. = 0015; 8:45 P.M. = 2045.

Exercises: Convert the following regular times to military time.

3:30 A.M.   =

10:00 P.M.   =

7:20 P.M.   =

4:10 P.M.   =

12:00 midnight   =

12:05 A.M.   =

LISTENING SKILLS

Listening is a basic skill utilized in interviewing, note taking and report writing. Good listeners use certain techniques to enhance their capacity to understand and gain information. Learning to be a good listener is similar to learning any new skill—it requires patience, time, practice, and some simple tools to help prepare yourself for the task ahead.

Prerequisites for Effective Listening

1. *Be mentally and physically alert.* Put yourself in listening position. Pay attention to each and every word.
2. *Attempt to understand the development of an idea.* Try to understand how the parts of a story or an argument are supported.
3. *Be receptive to what you hear.* Be objective and attempt to put aside personal feelings in order to hear the speaker's message before making your final evaluation.
4. *Be aware of the speaker's nonverbal cues.* Cues such as voice pitch, voice tone, and hand gestures are used to emphasize important material.
5. *Concentrate on the subject matter.* The average person speaks at a rate of 125 words per minute (wpm), understands and comprehends at a rate of 600 wpm, and thinks about subject matter at a rate of 60,000 wpm. Thus, there is time for the mind to review ideas very quickly. As you listen, your mind begins to establish relationships, classify events, and criticize the information it receives. Unfortunately, your mind may also wander and become distracted; this leads to inattentiveness and listening blocks.

Possible Blocks to Effective Listening

1. *Faking attention.* You miss information.
2. *Focusing on details or trying to memorize while listening.* You may lose the points and relationships between the overall theme and the details.
3. *Turning off.* If the material or the instructor is difficult or uninteresting to you, a more useful response would be to expend more energy, not less.

4. *Dismissal of the message.* You may reject ideas because of the speaker's mannerisms, delivery, or appearance.
5. *Distracting background noises.* Rather than be distracted by them, ignore them.

Mental Process Essential for Good Listening

1. *Think ahead.* Anticipate where the speaker might lead you.
2. *Weigh what you hear.* Be willing to hear the pros and cons of a case before giving your final analysis.
3. *Review periodically.* Watch for a pattern to emerge and attempt to put the pieces together for a clearer picture.
4. *Listen for the implied meaning.* If you need more information, ask for it.
5. *Become aware of emotional filters.* Know your biases, and recognize when they are impeding the flow of information.
6. *Use concentration skills.* Concentration, like any other skill, can be learned.

Although your role as a listener is to receive, good listening is actually an active process. As a listener, you cannot afford to just sit and absorb the material in a "playback" recording fashion. You need to move constantly back and forth between your awareness of the speaker's information, your processing of that information, and your reaction to the message and the speaker. Ask questions if you don't understand what is being reported. Clarify and summarize to make sure you understand.

NOTE TAKING

An important responsibility of a protection officer is note taking. *Notes* are brief notations of specific events and may serve as a basis for completing Daily Activity Reports (DARs) or incident reports. Notes also provide a greater degree of accuracy than memory.

Always carry your notebook with you. The front cover or first page of the book should include:

- your name and badge number
- your agency or company name
- dates

For every day on duty, the protection officer must keep notes on specific information. Write the date, shift, assignment, supervisor's name, weather conditions, and any specific events. Notes can be used to refresh your memory in the likely event that you are called to testify in court or asked to discuss a particular event with your supervisor or legal counsel. Do not write anything in your book that is not related to your job. Avoid profanity or other personal writing that could serve as an embarrassment to you.

The contents of your notes may vary, but usually would include the following:

1. Victim statements and description of condition.
2. Witness statements.
3. Notifications to significant persons (supervisors, police, medical assistance).
4. Evidence found at scene.
5. Your arrival time and departure time.
6. Weather conditions.

7. Lighting conditions.
8. Crime scene diagram.
9. How crime was committed (broken window, weapon used).
10. Suspect information (description).

When taking notes, allow space to add, change, or reorganize information. Before preparing an incident report, make sure that your facts are organized in some type of chronological order. In some agencies, tape recorders and cameras may be used to assist in documenting an event. Check with your supervisor before using any recording devices or cameras.

Daily Activity Reports

Protection officers often have the responsibility of recording events during their shifts. These reports are known as *daily activity reports* (DARs) or post logs. The purpose of a DAR is to document calls for service, unusual incidents, or patrol(post) duties. These reports provide information regarding your time and assignments.

A DAR should include the following information: date, name of reporting officer(s), time that the call or assignment was initiated and ended, source of entry or how call was received (radio dispatch), location of incident or function, disposition of duty, and reference to other reports if any. Figure 4–1 is an example of a DAR in use by airport safety forces. In Item 1 for example, the officer received a dispatch at 1610 Hours. At 1650 Hours, the assignment was completed, representing a 40-minute period. The location of the assignment was noted as the administration building. The disposition was a bank run to First National Bank, the contact person was Mr. Adams. There were no further reports. In Item 2, the officer was patrolling the north parking lot. There were no unusual events reported in the lot. In Item 3, the officer found an open door at the east warehouse. The officer discussed that Sam Jones, an employee, was working late. In Item 4, a visitor reported that the front windshield of her vehicle was smashed. An incident report was taken (see Figure 4–2).

The notes you take during your shift will serve to write your DAR. In some situations, you may carry a DAR and write your assignments as you complete them. The method you use will be determined by your agency and type of assignment.

In summary, a DAR should include the following information:

1. The time the assignment was initiated or received.
2. The time the assignment was completed.
3. Total time or duration of assignment.
4. The source of the assignment (radio dispatched or officer initiated).
5. The location of the assignment (address or specific property location).
6. Disposition. Indicate what happened or who was involved. Include any special remarks.
7. Additional reports or citations. Indicate if there were any follow-up reports (incident reports, arrests).

WRITING THE INCIDENT REPORT

The *incident report* is a narrative writing of a specific event. The incident may be a reported crime, safety violation, or injury. Whatever the event, an incident

OFFICER'S DAILY LOG

PAGE 1 OF 1   DETAIL TIME 1600 HRS.

DAY MONDAY   DAY 5 - 21 - 94

RADIO # 2351   BLANKET X

VEHICLE # 621   FLARES X   CAMERA X

MILEAGE IN 3715   FIRE EXTIN. X

OUT 3750

TOTAL 35   SPARE X

OFFICER(S) Ralph Davis   WATCH 1600 - 2400   ASSIGN _____

MISCELLANEOUS

RADIO CALLS	____
OBSERVED	____
OTHER	____

REPORTS

TRAFFIC ACCIDENT	____
INCIDENT	____
OTHER REPORTS	____
TOTAL	____

MISCELLANEOUS INFO

ITEM NUMBER	TIME RECEIVED	TIME ENDED	TOTAL TIME	SOURCE	LOCATION	NAMES, ADDRESSES, DISPOSITION, REMARKS, ETC.	DR#, A# OR CITATION #
1	1610	1650	40	Radio	Administration	Bank run, Mr. Adams	none
2	1650	1715	25	Ofcr	North Parking Lot	Patrolled lot, no unusual events	none
3	1715	1730	15	Ofcr	East Warehouse	Open door, employee working late, Sam Jones	none
4	1815	1845	30	Radio	South Parking Lot	Broken windshield, Mr. Anderson	see report

FIGURE 4-1

SECURITY DEPARTMENT INCIDENT REPORT			Report No. 920012	
Incident Type Vandalism of Vehicle				
Date Occurred 060594	Time Occurred 1825 Hrs	Location Occurred North Parking Lot		
Date Received 060594	Time Received 2400 Hrs	Officer(s) – List Writer of Report First Henry James	Approved By J. Jones	
Code: (V)ICTIM (C)OMPLAINT (S)UBJECT (E)MPLOYEE (W)ITNESS (M)ENTIONED				
Name	Address or Location/Department/Supervisor		Telephone or Extension	
(V) Vickie Smith	1624 Alta St., Pasadena, WA 80088		(555) 567-9832	
(W) Bill Johnson	22176 Tyrone Ave., Van Nuys, WA 80083		(555) 554-4485	

CODE: (L)OST (S)TOLEN (D)AMAGED (R)ECOVERED (N)ARRATIVE

DESCRIPTION, MAKE, SERIAL #, MODEL #   1990 Ford Thunderbird   VALUE

1. **Source of Activity**: I was dispatched to the north parking lot at 1825 Hours.

2. **Response:** Upon arriving at 1830 Hours, I met Vicky Smith standing next to her vehicle. The front windshield was smashed. Smith stated that she parked her vehicle at about 1620 Hours. She returned at approximately about 1815 Hours and noticed that the front windshield was smashed out! A large rock was found inside her vehicle. Witness Johnson parked his vehicle in the same lot at approximately 1730 Hours. He observed an unidentified male caucasian loitering in the lot near Smith's vehicle. The suspect was wearing blue jeans and green T-shirt.

No further description was reported. Suspect left in an unknown direction.

3. **Actions:** I gave Smith an information card indicating the report number.

I advised her to report the incident to her insurance company and local police. At Smith's request, her vehicle was towed by ABC Towing. Photographs were taken of Smith's vehicle and filed with the report.

NOTIFICATIONS MADE DATE/TIME	SECURITY MANAGER	DIRECTOR OF SECURITY	V.P.	EXECUTIVE V.P
INVESTIGATION NEEDED (Y/N):	FINAL DISPOSITION: __CLEARED __UNFOUNDED __INFORMATION ONLY	DATE OF DISPOSITION:	INVESTIGATOR:	

FIGURE 4-2

report may be needed to assist management or law enforcement in subsequent investigations. Incident reports are records; they can be used for future crime and safety planning. They can also be brought into court to verify or document a particular incident. It cannot be stressed enough how important it is to prepare these reports accurately and neatly. Whenever possible, these reports should be typed. However, if they are written, they should be printed clearly in pencil.

Incident reports must follow a basic format. The rule is to follow a chronological order of events. This means to write the report as you would write or tell a short story. Begin with how the incident was reported and what happened after you arrived at the scene. Conclude with what action was taken. The following format is recommended in writing an incident report. The report is written in three parts (Figure 4–2).

1. *Source of Activity*: The first part describes how the report was received. Usually, you will be dispatched by phone or radio to the incident. However, you may observe an event while on patrol. This part is usually short (one or two sentences).

   Example: At 1500 Hours I received a radio dispatch to go to the emergency room.

   Example: At 1500 Hours I observed a man down in the N/W corner of the South Garfield lot.

2. *Response*: This is the second part of the incident report. This part is usually the longest. It describes what happened upon your arrival at the scene. It includes statements provided by any witnesses or victims. This part is written in the order that you investigated the incident.

   Example: At 1515 Hours, I arrived at the emergency room. I was met by Dr. Bill Jones and E/R Nurse Jane Smith. They stated suspect Brown was being treated for a gunshot wound. Brown was combative and threatened to kill Smith and Jones if the police were notified. Brown was being restrained by orderlies John Johnson and Jim Anderson. Brown stated: "I will shoot this place up if the cops are called." Brown did not state how he arrived at the hospital.

3. *Actions*: This is the final part of the incident report. This part describes what was done. It provides additional information regarding the disposition of the incident or actions you took, including notification and other investigations.

   Example: Suspect Brown was shot in the left shoulder by unknown person. Brown had no personal identification. Brown did not provide any information regarding the source of his injury. The police were notified at 1545 Hours by Nurse Smith. Officers Green and White arrived at 1550 Hours and took a report (#992-4315). Brown was treated and transported to County Hospital. No further action was taken.

If observing an event taking place, include your observations and action.

Example:
Source of Activity: At 2045 Hours, I observed the rear stockroom door of Building C open.
Response: I checked the area for employees, no one was working. The outside lights were off. A vehicle, license # 25673, was parked approximately 30 yards from the door. I notified dispatch and was told to stand by and wait for assistance.
Actions: At 2055 Hours, Officer Smith arrived to assist in checking the stockroom. The search revealed employee John Mitchell taking inventory.

Mitchell's supervisor (Sam Adams) was notified and verified Mitchell's authority to be in the stockroom. Adams was advised to notify the protection office if employees are to be working after their regular shifts. No further action taken.

Remember the basic information needed in a report. Figure 4–3 summarizes the information needed in an incident report (Cox and Brown, 1992). Further guidelines in writing narrative reports:

1. Write in the first person. *Right*: I observed. I saw. *Wrong*: Officer observed.
2. Use an active voice. *Right*: Bill hit him. *Wrong*: He was hit by Bill.
3. Keep sentences short.
4. Use block lettering and print neatly. Do not write in longhand.
5. Avoid slang or jargon.
6. Use only recognized abbreviations.

BASICS	SPECIFICS	FURTHER DETAILS
WHO was the:	Complainant___ Victim___ Patient Witness___ Visitor___ Guest___ Shopper___ Employee___ Traveler___ Customer___ Doctor___ Nurse___	Description: name, sex, address, phone (home & business), occupation, physical, facial, clothing, ss#, license
WHAT started your action:	Foreseeable problem___ , describe___ Location___ Danger___ Remedy___	Precise details Repair or notification to person responsible
WHAT kind of problem:	Lost___ Found___ Criminal___ Items listed___ Non-routine___ Trespassing___ Accident___ , type___ Person___ Witness___	Precise value claimed by owner Estimate Insurance
WHEN:	Notified___ Viewed___ Assistance given___	Called maintenance Called police, arrival Called main office Called ambulance, arrival
WHERE:	Hospital___ Hotel___ Business___ Museum___ Other___ Department___ Room___ Hall___ Elevator___ Office___ Parking lot___ Street___ Restroom___ Swimming pool___ Other___	Exact address and location of specific area
HOW:	Accidental___ Presumed cause___ Witnesses___ Injuries___ Damage___ Help requested___ Help given___ Intentional___ Criminal___ Rape___ Drugs involved___ Burglary___ Robbery___ Unattended death___	Statements Notifications Police called Precise list of items Value

F I G U R E   4 – 3   Information typically found in an incident report (private security).

7. Complete all boxes on the report form. If a particular box is not necessary to complete, indicate N/A (not applicable).

8. Write your reports in a quiet place. Avoid distractions.

9. In writing a report, do not add anything. Stick to the facts as they are told to you. Do not give opinions unless asked!!

10. Do not write a report for someone else. If another officer observes an event, that officer should write the report.

11. Use common everyday words. For example, use *took* rather than *absconded*.

12. Check your spelling and use a dictionary.

13. Avoid wordiness. *Right*: The officer thought the suspect was ill. *Wrong*: The officer was of the basic opinion that the suspect was not feeling well.

14. Remember, reports are permanent records.

MISCELLANEOUS REPORTS

There are a number of miscellaneous reports in use. In cases of accidents or injuries, you may be required to complete a First Aid or Injury Report (Figure 4–4). If you make an arrest you will be asked to complete an arrest statement form (Figure 4–5). This form is more common for protection assignments such as shoplifting details. If assigned to parking control, the protection officer may be required to enter the time, license number, and description of all vehicles entering the property (Figure 4–6). Some protection officers are given the authority to write parking citations. The citations should be completed neatly and accurately (Figure 4–7).

Conclusion   The importance of writing clear and concise reports cannot be stressed enough. Improper reports, or reports lacking substance are harmful to the company and the officer. The ability to write quality reports requires an understanding of basic grammar and spelling. Make a practice of using notebooks or recorders when conducting patrol duties. Do not hesitate to record any unusual or suspicious events. It is also a good idea to review your reports with another officer or supervisor before final submission.

Learning   Identify the misspelled word or words in each of the following statements. Use a
Exercises   dictionary.

1. The officer believes the suspect was under surveillance.
2. The sergant was transferee.
3. Their are to officers assigned two the front gate.
4. The employee was suspected of embezzlement.
5. In case of a emergency, contact the lietenant.
6. Earthwuakes can have a desastrous affect on a city.
7. The investigation revealed that the acident was caused by human error.
8. Company personal are admitted after 0600 hours.
9. It was the officers belief that the burglay ocurred after midnight.
10. The supervisor disciplined the officer for being absent.

    The following sentences were taken from actual reports. How would you rewrite them?

1. It isn't clear whether she has slept with John, but it is clear that he has come in her bedroom window at times.

2. Couple is breaking up home, friends helping.

3. Man has diabetes and is insulated twice a day.

4. Applicant took job as a janitor in home for working girls—lasted three (3) weeks.

5. Man recently had operation, but is unable to hold any position, it seems.

6. I am writing the Welfare Department to say that my baby was born two years old. When do I get my money?

7. Mrs. Jones has not had any clothes for a year, and has been visited regularly by the preacher.

8. I cannot get sick pay. I have six children. Can you tell me why?

9. I am glad to report that my husband who is missing is dead.

10. This is my eighth child. What are you going to do about it.

11. Please find out for certain if my husband is dead. The man I was living with can't eat or do anything until he knows.

12. My husband got his project cut off two weeks ago, and I haven't had any relief since.

References   ADAMS, THOMAS F., *Police Field Operations,* Second Edition. Englewood Cliffs: N.J., Prentice Hall, 1990, pp. 127–129.

COX, CLERIC R., AND BROWN, JERROLD G., *Report Writing For Criminal Justice Professionals.* Cincinnati, Ohio: Anderson Publishing, 1992.

HODGES, JOHN C. AND WHITTEN, MARY E. (1984) *Harbrace College Handbook* New York: Harcourt Brace Jovanovich Publishers, 1984.

FIRST AID REPORT

DESCRIPTION OF VICTIM

NAME __Amy May Belle__   DRIVER'S LICENSE NO. __PO55989__

ADDRESS __4544 E.ADAMS ROAD__   RES. PHONE __310/555-4455__   BUS. PHONE __213/555-8832__

CITY __Los Angeles__   STATES __CA__   ZIP __90007__

LOCATION OF OCCURRENCE __645 E. Figueroa St., Los Angeles, CA. 90014__

CONDITION OF LOCATION   DRY ☐   WET ☐   SEE PHOTO ☐   OTHER ☐ _____

DATE & TIME OCCURRED __5/24/94 1400 HRS__   DATE & TIME REPORTED __5/24/94 1430 HRS__

INFORMANT __Chester Gomez__

TIME FIRE DEPARTMENT NOTIFIED __N/A__   TIME ARRIVED __N/A__

TRANSPORTED TO __N/A__

WITNESS __HILARY JOHNSON__

STATEMENT OF VICTIM

    As I was walking down the hall, I slipped on a wet spot and fell. I hurt my wrist as I used my hand to break my fall.

                 AMB

(This report hould be initialed by the person making the statement.)

OFFICERS NARRATIVE

    Ms. Belle was taken to the floor nurse who treated her sprained wrist and gave her aspirin. Ms. Belle had fallen when she slipped on a wet spot on the first floor hall near the cafeteria.

REPORTING OFFICER	ID#	DATE
Eileen Romero	32132	5/24/94

APPROVED BY	ID#	DATE

FIGURE 4-4

```
                  CITIZEN ARREST STATEMENT

                                            Date _____

As a private person _____
                    (Last Name)    (First Name)    (Middle)

have arrested _____
                    (Last Name    (First Name)    (Middle)

for (violation) _____

(Code Section) _____ , A public offense committed in my presence

at (location) _____
                          (Address of Offense)

At (time) _____    (date) _____   I hereby demand that

police offerer(s) _____  transport said

_____ to the _____ Police Station.
      (Name of Suspect)

I agree to appear in the office of the City Prosecutor before

4:15 A.M. _____ to sign a complaint against the above
              (Date)

arrested person. I agree to cooperate fully at all stages of the

proceedings.


                              _____
                              Signature of person making arrest

                              _____
                                                        Address

                              _____
                                               Telephone number


     (Fill out in duplicate. Second copy to person making arrest.)
```

FIGURE 4-5

VEHICLE LOG

DATE: _____
TIME: _____

MAKE	MODEL	YEAR	COLOR	LICENSE PLATE #	LOCATION	COMMENTS

FIGURE 4 – 6

NOTICE OF
TRAFFIC VIOLATION

DATE 10/31/94	TIME 1459	**No. 007104**

VEH. LIC # 1 ABC123	MAKE VW	MODEL BUG

VIOLATION(S)
1. 08   2. 11   3.   4.

LOCATION
   g–1 parking structure

   Level 1   Section C

☐ CHECK IF VEHICLE HAS BEEN IMMOBIILIZED

☐ CHECK IF VEHICLE HAS BEEN TOWED

YOUR VEHICLE IS PARKED IN VIOLATION OF THE RULE
CODED ABOVE. IF VEHICLE HAS BEEN IMMOBILIZED
PROCEED TO SECURITY AND INFORMATION DESK.

DESCRIPTION OF VIOLATION LISTED ON REVERSE

COMMENTS: *All employees are requested to park on the roof level*

SECURITY OFFICE PHONE NUMBER
      246-2409

SECURITY OFFICER *Officer Henry*	BADGE NO. # 35

MANAGEMENT OFFICE NUMBER
      240-9481

FIGURE 4–7

Interpersonal Communications

After studying this chapter, you should be able to explain the following:

The importance of effective
    communication
Sources of stress
Stress and personal health
Nonverbal communications
Empathy
Communicating professional
    language

Verbal judo techniques
Developmental disabilities
Mental illness
Thought and mood disorders
Mental retardation
Grave disability

INTRODUCTION

Crucial to the role of private protection is the ability to communicate effectively with the public as well as one's peers and superiors. *Communication* is the process of sending and receiving information and requires verbal as well as listening skills. Verbal communication, like written communication, demands that the sender express thoughts in a clear and meaningful manner. In interpersonal communications, information is relayed between two or more persons. This usually means face-to-face contact where body language is also observed. The manner in which a person communicates is just as important as what is communicated.

This chapter will discuss the importance of interpersonal communications as well as some strategies on how to deal with persons experiencing stress and anger. The protection officer is often put into the position of exerting authority over others. At times, some persons may resent or challenge this authority. Simply telling someone that parking is prohibited in a certain area may arouse expressions of anger and frustration. How a protection officer responds to these challenges can make the difference between a successful or disastrous outcome.

STRESS

There has been much written about stress. In today's society, it is hard to imagine life without some form of stress. Stress can be caused by events happening on the job, in our personal lives, or as a result of social changes. Whatever the cause, many people at one time or another have experienced stress. The situation that protection officers, like police officers, will have to deal with very frequently is another person's stress. Although protection officers may have their own problems, they are still expected to act professionally toward others. The field of private protection has a number of stresses such as shift changes, negative public perception, and low pay.

Experiencing stress does not always have to be negative. We sometimes undergo stress while attempting to be happy. Studying for a promotional exam, buying a new car, starting a different job, or getting married can be sources of stress. What we have to be concerned about is negative stress or stress that causes illness or unhappiness. Table 5–1 identifies the fifteen most common stress warning signs. While the table relates to policing, the information may be useful for protection officers as well.

Negative stress can cause unhappiness, illness, depression and other types of behavioral problems. Unless these problems are dealt with in a positive manner, life can be unpleasant. Your life style, along with other considerations, has a significant impact on your mental and physical health. Take a few minutes to complete the following exercise on calculating life expectancy shown in Table 5–2. What conclusions can be drawn from this test?

Table 5–3 tells what percentage of the population you will outlive, providing you make it to the specified age.

NONVERBAL COMMUNICATION

A form of communication often overlooked is nonverbal communication. All of us have experienced situations where others have given us nonverbal signs in the form of gestures and facial expressions. Placing one's hands on the hips may suggest authoritarianism. A person who crosses his or her arms while listening to another may appear judgmental or confrontational. Failing to look someone in the eye while

TABLE 5–1
Fifteen Most Prevalent Stress Warning Signs

Warning Signs	Examples
1. Sudden changes in behavior. Usually directly opposite to usual behavior.	From cheerful and optimistic to gloomy and pessimistic.
2. More gradual change in behavior but in a way that points to deterioration of the individual.	Gradually becoming slow and lethargic, possibly with increasing depression and sullen behavior.
3. Erratic work habits.	Coming to work late, leaving early, abusing overtime.
4. Increased sick time due to minor problems.	Headaches, colds, stomach aches, etc.
5. Inability to maintain a train of thought.	Rambling conversation, difficulty in sticking to a specific subject.
6. Excessive worrying.	Worrying about one thing to the exclusion of any others.
7. Grandiose behavior.	Preoccupation with religion, politics, etc.
8. Excessive use of alcohol and or drugs.	Obvious hangover, disinterest in appearance, talk about drinking prowess.
9. Fatigue.	Lethargy, sleeping on job.
10. Peer complaints.	Others refuse to work with him or her.
11. Excessive complaints (negative citizen contact).	Caustic and abusive in relating to citizens.
12. Consistency in complaint pattern.	Picks on specific groups of people (youth, blacks, etc.)
13. Sexual promiscuity.	Going after everything all of the time on or off duty.
14. Excessive accidents and/or injuries.	Not being attentive to driving, handling prisoners, etc.
15. Manipulation of fellow officers and citizens.	Using others to achieve ends without caring for their welfare.

Source: Territo, Leonard, and Harold J. Vetter, *Stress and Police Personnel*, p. 6. Boston: Allyn and Bacon, Inc., 1981.

communicating may suggest insecurity, shame, or deceit. If the nonverbal message conflicts with the verbal message, people may tend to believe the nonverbal message.

If an officer frowns while agreeing to work overtime, the nonverbal communications signal a reluctance to work in spite of the verbal agreement. There are generally four features of nonverbal communication:

1. *Gestures.* These include posture and movements of the body or parts of the body.
2. *Facial expressions.* The face shows feeling; for example, raised eyebrows suggest a surprise. Frowning suggests unhappiness or disapproval.
3. *Physical appearance.* Manner of dress and fitness level provide clues. A sharp appearance tends to make others think you're sharp. The appearance of a professional-looking uniform cannot be overstated. An officer who is overweight and wears a sloppy uniform may not get the required respect.
4. *Touching.* In some cultures, touching is accepted as part of the process of communication in social situations. Placing a hand on someone's shoulder may be offensive to some but accepted by others. This form of touching suggests an eagerness to communicate or to get someone's attention.

T A B L E 5 – 2
Life Expectancy Test

If you are between twenty and sixty-five and reasonably healthy, this test provides a life insurance company's eye view of the future.

1. Start with seventy-two.
2. *Gender.* If you are male, subtract three. If you are female, add four. That's right, there's a seven-year spread between the sexes.
3. *Life-style.*
   (a) If you live in an urban area with a population over two million, subtract two. If you live in a town under ten thousand, or on a farm, add two. City life means pollution, tension.
   (b) If you work behind a desk, subtract three. If your work requires regular, heavy physical labor, add three.
   (c) If you exercise strenuously (tennis, running, swimming, etc.) five times a week for at least a half-hour, add two.
   (d) If you live with a spouse or friend, add five. If not, subtract one for every ten years alone since age twenty-five. People together eat better, take care of each other, are less depressed.
   (e) Sleep more than ten hours each night? Subtract four. Excessive sleep is a sign of depression, circulation diseases.
   (f) Are you intense, aggressive, easily angered? Subtract three. Are you easygoing, relaxed, a follower? Add three.
   (g) Are you happy? Add one. Unhappy? Subtract two.
   (h) Have you had a speeding ticket in the last year? Subtract one. Accidents are the fourth-largest cause of death; first, in young adults.
4. *Success.*
   (a) Earn over fifty thousand dollars a year? Subtract two. Wealth breeds high living, tension.
   (b) If you finished college, add one. If you have a graduate or professional degree, add two more. Education seems to lead to moderation; at least that's the theory.
   (c) If you are sixty-five or over and still working, add three. Retirement kills.
5. *Heredity.*
   (a) If any grandparent lived to eighty-five, add two. If all four grandparents lived to eighty, add six.
   (b) If either parent died of stroke or heart attack before the age of fifty, subtract four.
   (c) If any parent, brother, or sister under fifty has (or had) cancer, a heart condition, or has had diabetes since childhood, subtract three.
6. *Health.*
   (a) Smoke more than two packs a day? Subtract eight. One to two packs a day? Subtract six. One-half to one? Subtract three.
   (b) Drink the equivalent of a quarter-bottle of liquor a day? Subtract one.
   (c) Overweight by fifty pounds or more? Subtract eight. By thirty to fifty pounds? Subtract four. By ten to thirty pounds? Subtract two.
   (d) Men over forty, if you have annual checkups, add two. Women, if you see a gynecologist once a year, add two.
7. *Age adjustment.* Between thirty and forty Add two. Between forty and fifty? Add three. Between fifty and seventy Add four. Over seventy Add five.

The modern protection officer should understand the importance of nonverbal communication. When taking a report or interviewing someone who has been victimized, notice whether the nonverbal signs are the same as verbal messages. Describe the nonverbal messages being given in Figure 5–1. People often make judgments about others in a matter of minutes. A protection officer's nonverbal image must be positive.

FIGURE 5-1

VERBAL COMMUNICATIONS

The words the protection officer speaks and how they are spoken can make the difference between improving a situation or making it worse. Effective communication requires the protection officer to listen closely to what is being said. Listening with purpose is just as important as speaking with purpose. Good listening requires the protection officer to put himself or herself in the position of the speaker. This sharing of another's feelings is called *empathy*. Protection officers confronting stressful encounters will find this tactic particularly useful.

Empathy should not be confused with sympathy. If a person is sad due to the death of a loved one, we would show sympathy by being sad. Empathy is the

TABLE 5-3
How to Calculate Your Life Expectancy

Age	Men	Women
60	26%	15%
65	36%	20%
70	48%	30%
75	61%	39%
80	75%	53%
85	8%	70%
90	96%	88%
95	99%	97%
100	99.9%	99.6%

Source: Territo, Leonard, and Harold J. Vetter. *Stress and Police Personnel*, pp. 60–61. Boston: Allyn and Bacon, Inc., 1981.

ability to show others that we understand what they are saying about their feelings. If a protection officer is upset because his supervisor accuses him of being neglectful on his post one day, a response with empathy would be as follows:

"It's upsetting to be thought of as not doing your job. Although you are referring to one incident, it sounds like maybe you are questioning your ability to do your job, or relations with your supervisor."

In this type of response, the listening officer is allowing the person to express himself or herself further. In other words, the complaining protection officer may respond by saying

"I guess maybe you are right. Things haven't been going too well…"

In empathy, an attempt is made to let the person talk and express feelings. Empathy means more than telling a person who has undergone a personal loss

"That's too bad. Life is tough."

A better response in this situation would be as follows:

"It is upsetting to be out of a job. You must feel lost or unsure about what to do."

This initial approach will probably get a response from the person. After establishing communications and sharing understanding, the protection officer can offer alternatives or suggestions.

As discussed, verbal encounters may occur during stressful situations. In fact, you may have to deal with angry persons daily. There are many times when a person may be prone to anger. Some examples are:

1. Frustration or not being able to meet a goal.
2. Not feeling good about self or life accomplishments.
3. Immaturity or emotional instability.
4. A defensive tactic to keep others from getting too close.
5. A plea for help due to some personal sadness.

Understanding that people expressing anger may be doing so because of some personal problem can help the protection officer respond. An officer should not fall into the trap of getting angry because someone else is angry. Losing self-control suggests a weakness in the officer's ability to respond or help. Incidents which may cause a person to be angry can be minor or major. A business patron unable to find a parking space, an employee upset over company policy, a business dispute, a domestic argument, and an automobile accident are all examples of anger-producing situations to which a protection officer may have to respond.

In responding to an incident where a person or persons are expressing anger, attempt to apply the following guidelines:

1. *Know and understand your own responses to anger.* Anticipate comments that may make you angry.
   Example: "This is none of your business, officer." "Get lost!" "You're not a cop!"
2. *Remember why anger occurs.* Anger can come from a person who is unhappy. An employee upset over a work rule may be unhappy and unfulfilled in his or her personal life.

Example: "It seems like everyone is trying to control my life."

3. *Let the angry person talk.* Be a listener. Allow the person to relieve the anger, even if it is directed toward you. Allowing the person to talk will relieve frustration.

   Example: "You protection people are always trying to tell someone what to do. Well, let me tell you what I think…"

4. *Show nonverbally that you are listening.* As discussed, nonverbal signals are important indicators of what we feel or think at times. A verbal response such as, "I'd like to listen to what you have to say," or, "this must be very important to you," will be much more meaningful if the officer making the statement nods affirmatively, keeps good eye contact, and shows concern. Avoid negative nonverbal signs such as yawning and poor eye contact.

5. *When the angry person is ready to speak, respond to his or her feelings.* Show that you understand how and why the person feels the way they do.

   Example: "I can understand why you're angry over not getting a pay raise, especially after all the extra work you've had to do. These are difficult times and every little bit helps."

The protection officer is not expected to solve everyone's problem. However, by being a good listener and by showing a caring attitude, the protection officer can help to reduce a person's anger.

6. Avoid the use of first names, slang, or other personalized language.

   Examples: "Hey dude, what's happening?" "Go for it lady!" "How are you Babe?"

In both verbal and nonverbal communication, it is important to use professional language. When confronting clients, visitors and others, respect must be shown. Unless others invite informality, always begin the communication with a formal, courteous greeting.

   Examples: "Good morning Dr. Smith." "Good afternoon Mr. (Mrs.) Jones." "Good evening Sergeant Adams."

7. Regardless of a person's social standing, always treat the person with respect. A good rule of thumb is to treat others the way you would like to be treated! Regardless of age, position, gender, race, or nationality, all people deserve formal, respectful communication. Formal greetings such as *Sir, Ma'am, Mr., Ms., Miss,* and *Mrs.* are more effective than negative greetings such as *mister, lady, boy, girl,* or *hey brother.*

The manner in which a person is initially greeted has a major impact on the remainder of the communication. If the initial encounter is negative or insulting to a person, then any attempt to communicate further will be difficult. Encounters with angry persons may make it difficult to act professionally; however, it is important to remember that it is the situation or some other event that is causing the person to be angry. The professional protection officer must remain calm and continue to communicate with respect.

Example:

Situation: Angry customer demanding to see the company president about a defective product.

Customer: "I want to see the president about this lousy product."

Officer: "I am sorry sir, but the president is unavailable. If you'd leave your name and number a representative will contact you, or if you like I'll attempt to contact someone in customer relations."

Customer: "I want to see the president now."

Officer: "Sir, I know you're angry, but it is not possible to see the president at this time. If you'll wait, I'll attempt to contact a company representative who'll be more than happy to discuss your concerns."

In this situation, the officer must defuse the anger of the customer. If the officer would challenge or insult the customer (for example, "Look buddy, nobody gets by without my say so," or, "I don't like your tone of voice. Get out of here!"), the situation would likely get worse. As discussed later in this chapter, the skillful use of communication can do a lot to control or defuse an unpleasant situation.

Paraphrasing

Paraphrasing is a rewording of a person's message in an attempt to clarify meaning. It is a useful tactic in improving communication because it allows the person to correct you if you are wrong. Paraphrasing offers the following benefits:

1. You can interrupt someone and not generate resistance.
2. You take control of the encounter.
3. You "get it right" on the spot. We don't always hear well.
4. The other person can correct you if you have made an error. It makes the other person feel good, and is good for you.
5. It makes the other person a better listener. No one will listen harder than to his or her own point of view.
6. It creates empathy. The other person will believe you are trying to understand.
7. It often makes the other person modify his or her initial statements (become more reasonable) because he or she gets to examine the meaning in different words and tones.
8. It overcomes sonic intention. People often think they have said something to you because they have "heard" themselves say it in their mind. When you paraphrase, they hear what you have heard, not what they think they have said.
9. It can clarify for those who may be standing nearby.
10. It prevents metaphrase—the use of the skewed phrase. The good paraphraser will never "put words into the other person's mouth."
11. Whenever you give directions to others, insist that they paraphrase back to you. Eight out of ten people misunderstand the point of a verbal exchange.
12. Whenever you take directions, paraphrase back to ensure you heard and interpreted correctly. The other person may not have said what he or she intended. Paraphrasing keeps you from making errors; it makes you efficient and effective.
13. It reinforces your own memory. The mind remembers what the mind does. Your reports, written or oral, will be more concise and accurate.
14. It generates the fair-play response. You have listened and made an effort to understand the other person. The other person is then inclined to do the same for you.

There are a number of situations in which conflicts can occur. For example, protection officers may need to intervene in work place disputes between employees. Business invitees may also get involved in disputes. The use of words to resolve conflict is preferred over physical force. Many police officers are undergoing training in the art of verbal judo. *Verbal judo* is the mastery of communication by redirecting behavior with appropriate words.

It is too easy to use force or violence against another. Your role as a protection professional is to be in contact with yourself, your employer, and the public; and to redirect behavior, empathize, influence others, and seek voluntary compliance. The key to understanding verbal judo is to recognize that words are more effective than physical violence.

A person's anger and verbal attacks most often are the result of an unpleasant experience or other frustration. While the protection officer may be the target of such abuse, the officer is usually not the source of the problem. While it is difficult to stand by while someone directs profanity or engages in name calling, it will only make matters worse if the protection officer retaliates with negative words or profanity. This is referred to as *escalation of conflict*, which can result in the use of force and physical injury. If an angry person tells a protection officer to perform an act of sexual intercourse upon himself, or accuses the officer of having sexual relations with a close family member, the question the officer has to resolve first is whether these allegations are true. Since these insults are untrue and ridiculous, why would a protection officer take offense to the point of using force. Using force or returning the name calling and profanity is a sign that the officer is losing control of the situation. In dealing with conflict, a protection officer must control the situation, and not allow the situation to control the officer.

The question is frequently asked: "How do I respond to a person who insults me, my manhood (womanhood), my spouse, or my sexual orientation?" The following situation is an example of the incorrect and correct way of responding.

Situation: A distressed woman arrives at a hospital emergency room to see her husband who has been seriously injured in an automobile accident. Her husband is undergoing emergency surgery.

Woman:	*"I demand to see my husband! I want to be with him!"*
Protection Officer:	*"I'm sorry, but you'll have to wait in the lobby. He is undergoing surgery."*
Woman:	*"Look stupid, let me through!"*
Protection Officer:	*[improper response] "Look lady, you try and I'll have you arrested! Don't talk to me like that!"*
Protection Officer:	*[proper response] "I know you are concerned and I'm sorry about the accident. The medical staff is doing everything they can. It's best you remain here. If you want I'll have someone check and see how the operation is going and have someone speak with you."*

In summary, the following guidelines are important to remember in using verbal judo.

1. Never use words that rise most readily to your lips.
2. If it makes you feel good, it's probably no good.

3. Only show your professional face(s), never your personal feelings.
4. If you can't control yourself, you can't control others.
5. If you lose your sense of humor, you're dangerous to yourself and others.
6. The less ego you show, the more power you have over others.
7. The most dangerous weapon you have is your cocked tongue.
8. Never tell someone to calm down. Calm them down by your performance and tone of voice.
9. If you say the first thing that comes to mind when you are angry, you risk making a speech you'll live to regret.
10. When your mouth opens, your ears should also open (Thompson, 1992).

DEALING WITH MENTALLY DISTURBED PERSONS

In the course of the protection officer's work there will be situations involving persons with developmental disabilities or mental illness. Protection officers must be able to recognize and identify the signs of mental illness and developmental disabilities in order to safeguard their own safety, the safety of others, and the safety of the developmentally disabled or mentally ill person. The two types of mentally disturbed persons discussed in this section are the developmentally disabled and the mentally ill.

Developmental Disabilities

Developmental disabilities are present from birth and categorized as follows:

1. mental retardation
2. epilepsy
3. cerebral palsy
4. autism

Mental retardation is a disability characterized by lower-than-average intelligence, while epilepsy is a neurological disorder caused by brain damage. In epilepsy there are two types of seizures: 1) grand mal or tonic/clonic, and 2) petit mal or absence. These differences relate to the severity of the symptoms. Cerebral palsy is caused by damage to the area of the brain which controls motor function. Autism is characterized by severe behavioral and communication disorders.

Identifying behaviors.   The protection officer must be able to recognize the wide range of behaviors that are symptoms of persons with developmental disabilities.

Mentally retarded persons.   Mentally retarded persons will have a significantly lower-than-average intellectual function. For example, a 25-year-old mentally retarded person may only be functioning at a mental level of a person eight years old. These persons will usually appear physically normal and have difficulty speaking clearly. However, the following symptoms may be present:

- muscle control difficulty (in severe cases)
- Down's syndrome
- confused or disoriented appearance (in moderate to severe cases)
- self-endangering behaviors (in moderate-to-severe cases)

- inappropriate responses to situations
- purposeless repetitive behaviors
- deficits in common knowledge (for example, adult individual may not be able to count coins or tell time)

Epileptic persons.   Epileptic individuals suffer from a neurological disorder due to brain damage and may have seizures and convulsions. There is loss of memory after a seizure accompanied by staring or falling asleep. Also present may be small twitching movements, yet they appear physically normal. Persons suffering from epilepsy are usually on medication to control their seizures. After suffering seizures, the following symptoms may occur: muscle-control difficulty, slurring of speech, confusion or disorientation, and lethargy.

Cerebral palsy.   Persons with damage in the area of the brain that controls motor function will have muscle-control difficulty as well as a slurring of speech. However, these persons will usually have normal mental capacity. The following symptoms may be present: mental retardation, difficulty hearing or speaking, seizures or spasms, and drool.

Autism.   Autistic persons behave unpredictably and have difficulty forming relationships with other people. They have problems understanding or speaking and appear confused or disoriented. Autistic persons often exhibit purposeless repetitive behaviors and have deficits in common knowledge. The following symptoms may be present: self-endangering behaviors (for example, person may wander into traffic), lack of perception regarding danger, inappropriate responses to situations, fearfulness to the point of running away. Autistic persons may echo others' words and have an unusual fascination with inanimate objects.

Mental Illnesses

These generally occur later in life as a result of biological, social, or psychological factors. They are classified as follows:

1. *Thought disorders*. One of the most severe forms of mental illness.
2. *Mood disorders*.
3. *Substance abuse* This type of abuse leads to organic damage.

Identifying Behaviors.   The protection officer must learn to identify behaviors which are symptoms of mental illness. Mentally ill persons exhibit a wide range of behaviors depending on their illness or combination of illnesses. Look for and learn to recognize behaviors and/or moods in a person that appear to be inappropriate to the situation and see if the behavior of the person appears inflexible and/or impulsive.

Behaviors particularly symptomatic of mental illness generally include the following:

- *Thoughts of death and/or suicide*. These symptoms are often associated with depressed persons. Threats, suggestions, and attempts at suicide should always be taken seriously. Persons suffering thought disorders may be responding to internal voices which issue violent commands.
- *Impaired self care*. As a result of mental illness, persons may not feed, clothe, or shelter themselves.

- *Impulsive, erratic or bizarre behavior.* Behaviors to look for include head banging, self-mutilation, rigid and unusual postures, inappropriate nudity or sexual behavior, directing traffic, and running in or lying down in traffic.
- *Disorientation.* Mentally ill persons often are unaware of the time of day, place, or identity of self or others.

Thought disorders. Mentally ill persons suffering thought disorders typically exhibit the following behaviors: delusions, hallucinations, or disorganized speech patterns. An example of a *delusion* or false belief is a person who may believe she is being persecuted, attacked, harassed, cheated or conspired against. Another example might be a person who believes he is Jesus Christ or the devil or might have special powers. *Hallucinations* are false perceptions through any one of the five senses. A person may exhibit symptoms such as hearing voices, seeing visions, or feeling bugs crawling on the skin. Persons exhibiting an inability to concentrate or make logical thought connections are suffering *disorganized speech patterns.* For example, the individual may have a rapid flow of unrelated thoughts; exhibit unclear speech that does not communicate an idea; speak incoherently (words do not fit together); make up words; talk in rhymes without regard to meaning; repeat the same words and phrases, or fail to; or be slow to respond to simple questions or respond with blank stares.

Mood disorders. The following are behaviors typical of mentally ill persons suffering from mood disorders:

- *Irrational fear or sense of panic.* Examples include persons who experience intense anxiety or panic attacks or those who may experience blind flight or paralyzed immobility (inability to move).
- *Depression.* Symptoms of depression a person may exhibit include feelings of overwhelming hopelessness, guilt, despair, and worthlessness. Also the individual may be lethargic or suicidal.
- *Mania.* Persons in this emotional state may show feelings of extreme elation or excitement and may be highly active.

Situations Requiring Intervention. The behaviors described in this section (which are the result of a mental disorder and not a life-style choice) may require the protection officer to take appropriate action in situations where the person may be a danger to himself, a danger to others, or be gravely disabled. How will the protection officer know when a person is a danger to self, others, or gravely disabled?

Danger to self. This situation occurs under the following circumstances:

- When a person has indicated by words or actions an intent or plan to commit suicide or inflict bodily harm on self.
- The person's plans or means are available or within the person's ability to carry out.
- The person shows such gross neglect for personal safety that he or she receives or is at risk of receiving serious injury.

Danger to others. This situation occurs under the following circumstances:

- When a person has indicated by words or actions an intent to cause bodily harm to another person.

- The person's threats or intentions are specific as to the particular person to whom harm would be done.
- The person, although not focused on a particular individual, is agitated, angry, and appears explosive.
- The person is engaged in or intends to engage in acts or behavior of such an irrational, impulsive, or reckless nature (destruction or property or mis-use of a vehicle) as to put others directly in danger.
- The person's acts or words regarding an intent to cause harm to another person are based on, or caused by the person's mental state which indicates the need for psychiatric evaluation and treatment.

Gravely disabled. A person is considered gravely disabled if he or she shows difficulty maintaining the necessities of life.

- *Food*
  - person is malnourished and dehydrated
  - little or no food in the house and the person is unable to establish where or how meals are obtained
  - no realistic plan for obtaining food
  - has repeatedly indicated intention to no longer eat or believes food is poisoned
  - frequently obtains food from garbage cans or similar sources
  - repeatedly eats items not ordinarily considered fit for human consumption
  - has been losing substantial weight without reasonable explanation
- *Clothing*
  - repeatedly destroys clothing
  - regularly fails to wear clothing in keeping with prevailing climatic conditions
  - clothing is repeatedly grossly torn or dirty
  - has no realistic plan for obtaining needed clothing
- *Shelter*
  - frequently sleeps in abandoned buildings, doorways of buildings, near public thoroughfares, in prohibited areas, or in other than ordinary shelter
  - repeatedly ejected from living quarters by landlords because of behavioral problems
  - no realistic plan for obtaining shelter due to present mental state and does not appear to be able to care for self

Intervention Strategies. Attempting to help mentally ill persons requires caution. It is important that safety practices be observed. The following guidelines are suggested when responding:

1. Be sure that there are no injuries or other medical problems requiring immediate attention. Look for obvious signs of injury. Request medical assistance when appropriate.
2. Protect yourself against injury. Be sure that the person is not armed or otherwise in possession of a possible weapon (for example bottles). Watch their hands.
3. Contact the police if it appears that the person is of unsound mind or unable to care for himself.
4. In attempting to communicate, treat the person with respect. Do not insult or anger the person. Attempt to humor the person. Keep him or her talking.

Conclusion   Effective communication requires the ability to understand others. In dealing with conflict or stressful encounters, the professional officer must have a good command of verbal and nonverbal language. The ability to be a good listener is important since listening skills form the basis of action. Angry words used against an officer do not justify physical response. Likewise, using angry words against one who is insulting or angry will likely make matters worse. In other words, be smart, be calm, and use common sense.

Discussion Questions
1. What is the difference between positive and negative stress?
2. Identify as many examples as you can of stress associated with the role of private protection.
3. Discuss the importance of empathy in the process of communication.
4. How can nonverbal gestures conflict with spoken words?
5. Discuss the recommended approach when trying to communicate with someone who is angry.
6. Why is the art of paraphrasing considered an important tool in communication?
7. Discuss the importance of verbal judo in conflict situations.
8. What are some behaviors that would be indicative of persons who are mentally ill and/or developmentally disabled?

References

KRIER, BETH ANN, "Copping an Attitude: Ex-policeman Teaches Officers this Art of Verbal Judo-reading and Redirecting People," *Los Angeles Times*, May 8, 1990.

Managing Contacts with the Developmentally Disabled or Mentally Ill, Student Workbook. State of California Commission on Peace Officer Standards and Training, June 1990.

MYERS, GAIL E., AND MICHELE T. MYERS, *The Dynamics of Human Communication*. New York, NY: McGraw Hill, 1980, p. 221.

TERRITO, LEONARD, AND HAROLD J. VETTER, *Stress and Police Personnel*. Boston: Allyn and Bacon, Inc., 1981, pp. 6, 60–61.

THOMPSON, GEORGE, *"The Art of Paraphrasing."* Workshop Handout. Verbal Judo Institute, Tijeras, New Mexico, 1992.

Physical Security Measures

After studying this chapter, you should be able to explain the following:

Alarm systems	The four D's of protection
CCTV systems	Intrusion sensors
Three lines of property defense	Protective lighting
Perimeter protection	Locking devices

There is no substitute for a well-trained, alert protection force. In many businesses, physical protection measures must be used along with a qualified protection force. As a protection officer, you will be expected to be thoroughly familiar with your area of protection and to know the location of all protection measures in use. You don't need to be an expert on alarms, locks, or other physical protection measures, but you should know how to use the equipment. In the following pages, a more thorough discussion of these defenses is presented. Remember, access control is the primary goal of protection.

While it may not be possible to defend against all intrusions, the use of these measures will greatly improve the protection profile. A protection officer may be assigned the responsibility of inspecting these defenses to ensure they are adequate. A good defense also includes protection measures such as fencing, locks, alarms, closed-circuit television cameras, and lighting. Regardless of the measure used, the four Ds of protection are:

- Deter the intruder.
- Detect the intruder.
- Delay the intruder.
- Deny the intruder access.

Lines of Defense

There are different types of measures used to protect property from unlawful intrusion. Generally, the three primary lines of defense used to provide protection are:

1. Perimeter barriers (parking lots)
2. Structure protection (buildings)
3. Point or target protection (rooms or specific targets within a structure such as safes, vaults)

Perimeter Barriers.   Perimeter defenses (see Figure 6–1) are designed to keep unwanted persons out and to define the property boundaries. There are many types of perimeter barriers. Some are natural barriers such as rivers, cliffs, and canyons; others are man-made. The man-made barriers include fences, walls, and other structures. The type of perimeter defense will depend upon the prop-

LINES OF DEFENSE

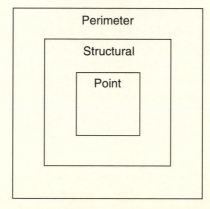

FIGURE 6–1

erty being protected. Some businesses may not have any barriers while others have several lines of perimeter defense.

If fencing is used as a perimeter barrier, it should be at least seven feet high, of the chain-link type. If barbed or razor wire is used, it should be at least one foot in height. An advantage of chain-link fencing is that it affords visibility for protection officers and passing police officers.

Structural Protection.   The second line of defense is building or structural protection. This protection involves the security of doors, windows, roof hatches, and sky lights.

According to insurance statistics, 90 percent of all building break-ins are through existing openings, 49 percent are through windows, and only about 3 percent are through roof openings. Since windows present special security problems, it may be wise for management to use iron or steel bars, or steel mesh on all openings larger than 96 square inches. This is important if the windows are close to the ground or near other structures. If bars are used, they must be inspected at regular intervals to ensure that they are secure.

Point or Target Protection.   This refers to the protection of specific objects or areas such as interior rooms, vaults, safes, and cabinets. In a hospital, for example, storage cabinets for narcotics require point protection.

Security Systems

Security systems refer to physical protection measures. These measures are locking devices, alarms, cameras, lighting, and fencing. These measures are used to facilitate the overall security plan.

Locking Devices.   Associated with building protection are a number of locking systems. Locks are only as good as the doors. For example, the most secure locks are useless with hollow wood doors. There are several types of locking devices. One type is the *warded lock.* This is an old locking device which utilizes a key with specially cut grooves (Figure 6–2). It is also referred to as a *skeleton key.* These types of locking devices do not meet any security specifications; therefore, these types of locks are rarely used.

Disc-tumbler locks, also referred to as *wafer tumblers*, are key operated and designed primarily for automobiles, desks, cabinets, and some padlocks (Figure 6–3). These locks provide more security than warded locks, but are still easily compromised.

Pin-tumbler locks are used by commercial and residential structures and they are more secure than the warded or disc-tumbler locks. In this type of lock, the key is inserted into a plug to turn open the lock (Figure 6–4). Pin-tumbler locks are used on doors, windows, and padlocks. Although these locks are more secure than the warded or disc-tumbler lock, they are prone to being "shimmied" open. To prevent shimmying, a dead latch is used. *Dead bolt locks* are often used in residential buildings such as apartments.

Combination locks are used in padlocks, vaults, and fence doors. As a general rule, combination locks are more secure than traditional key locks. The normal

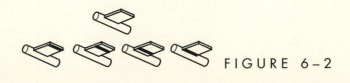

FIGURE 6–2

combination lock has a three-number combination, thus, if the lock has 60 numbers, then the number of possible combinations is 60^3 or 216,000 possible combinations. Most intruders will not take the time to manipulate the combinations. Safes and vaults are normally defeated by drills or explosives.

Electromechanical locks and *electromagnetic locks* are also used in commercial settings, and operate through electricity. Obviously, if power shortages occur, backup systems are necessary; otherwise, the lock will operate in a fail-safe cate-

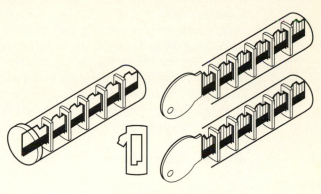

FIGURE 6-3

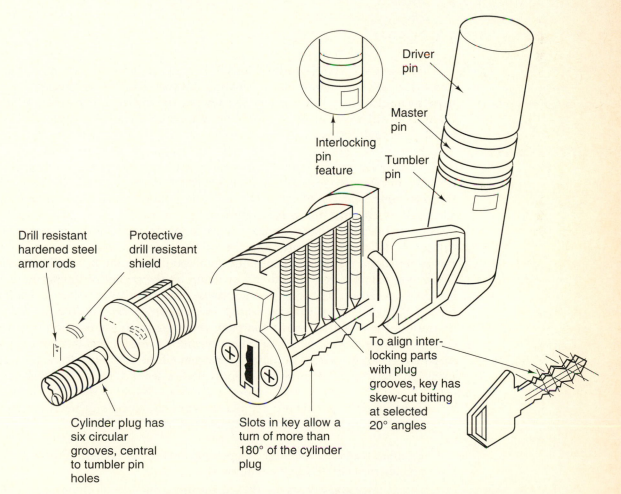

FIGURE 6-4   Pin tumbler cylinder lock with interlocking pins.

gory (unlocking automatically and opening for emergencies). Electromagnetic locks use electricity to power an electromagnetic door jamb, which exerts a force of 1500 pounds or more on a plate in the door, keeping it closed.

One of the primary problems of a locking system using keys is key control. To prevent too many keys from floating around they should be regulated to assure that they do not leave the premises. Master keys and combinations should be kept in secure places. It is recommended that periodic rekeying of the business be undertaken (at least annually or after significant employee turnover). However, the extent or need for rekeying will of course be a management decision. Many hotels and other commercial establishments are utilizing *plastic-card key systems*. These plastic cards contain coded information that is read by the system. When the card is inserted into the door, transmissions between the sensor and the card result in a code authorizing entry. Obviously, when employees or users terminate use, the cards must be collected or otherwise decoded. These systems eliminate the use of keys and offer more security since each card is coded differently making it difficult to bypass the system.

Alarms.   Alarms, or intrusive detecting systems, are available to all three lines of defense: perimeter, structural, and point or object protection. The use of alarms is dependent upon the type of business and protection needs. Alarms consist of three basic parts:

1. Sensor
2. Circuit
3. Signal

The *sensor* is the triggering device that detects an intrusion or condition. The *circuit* is the communication activated by the sensor. It receives and transmits the information. The *signal* alerts someone to respond and can be audible or silent. Many alarm systems transmit over phone lines to alert a response. Figure 6–5 depicts several types of sensor systems.

Types of Alarms.   Alarm systems and responses can be of several types. The first is referred to as a *local alarm system*. This type of alarm is audible and requires that someone hear the alarm and notify the police or fire department. These alarms are ineffective unless someone hears them. However, the alarm noise can discourage intruders; they often leave the area rather than risk the chance of being caught. Local alarms are prone to false alarm due to weather or careless use. *Proprietary alarm systems* are constantly monitored and may be audible or silent. These systems are often owned and operated by the owner of the property. The business protection officer would normally respond to these alarms. The third type of alarm is referred to as the *central station alarm system*. This type of system is monitored by an alarm/guard company removed from the location. When the central station receives the alarm from a business or residence, the company will dispatch officers or call the location to determine if the alarm is false. If there is an intrusion, the responding officer will notify the owner and/or call the police.

Types of Sensors.   There are many types of sensors in use by business today. Sensors can be divided into several categories.

1. *Electromechanical.* Operated by opening, closing, or changing the balance of an electrical circuit (doors, windows).
2. *Sound wave and microwave.* Based on the principle of changes in echoed waves as in radar detection.

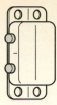

Magnetic contacts are attached to doors, windows, etc., so that when the door is opened the contacts are separated.

Pressure mats are usually placed under carpets and react to pressure from footsteps.

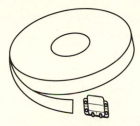

Foil is attached to glass or other surfaces and breaks when the surface is broken.

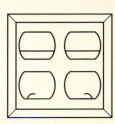

Photoelectric beams cast an invisible infrared light beam across doorways, etc., and react when the beam is interrupted.

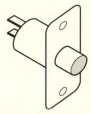

Plunger contacts operate in the same way as the light switch on a refrigerator door.

Motion detectors transmit and receive patterns of ultrasonic or microwave radiation. The pattern is changed when a person enters it, causing the detector to react.

Electric and magnetic field devices create stable fields close to specific targets such as safes and react when a person or object enters the field.

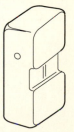

Vibration detectors are attached to surfaces and react to vibrations caused by attempts to break through the surface.

FIGURE 6-5

3. *Capacitance.* Used in protection of specific objects; touching of an object (for example, a safe) causes an alarm.
4. *Vibration.* Detects any movement of any object protected, such as in rooms.
5. *Audio.* Triggered by sounds picked up by acoustic microphones.
6. *Light.* Often referred to as *laser beams*. Based on light beams projected across an area.
7. *Holdup Alarms.* Sometimes referred to as *panic alarms*; activated by buttons, foot rails, or cameras.

Alarms are activated under any of the following conditions:

1. Penetration of the area being protected
2. Power failure
3. Opening, shorting, or grounding of the circuits
4. Failure or aging of a sensor
5. Tampering with the system

In responding to an alarm, never assume that it is false. While most alarms are false, there is a slight possibility that an intruder may be present. Use caution when entering a structure. Avoid entering alone since the intruder may be armed, or there may be more than one intruder. Look for open doors, windows, suspicious vehicles, and persons. Check the area surrounding the structure. If the alarm is false, report the reason for its activation. As indicated, there may be a number of reasons why an alarm has been activated. Table 6–1 describes the various types of intrusion devices as well as the advantages and disadvantages of each. The use of metallic foil on windows (Number 4), for example, is common practice for some businesses. The disadvantage of this device is that replacing windows with foil can be expensive, especially in neighborhoods with a high rate of vandalism.

CCTV Systems.   Closed-circuit television systems (CCTV) allow the protection officer to view several locations. CCTV cameras are normally placed in locations requiring special surveillance. In a parking lot, for example, cameras may be mounted on poles, or on roofs or sides of buildings in and around the lot. The placement of the camera takes planning. Placement is determined by the visibility required, with the goal of providing the needed surveillance for the entire lot. Some cameras may overlap surveillance, making it possible to view other camera locations. This is done to assure that other cameras are not obstructed or damaged.

The CCTV monitors are located in the protection office or other designated locations. Usually, the monitoring officer will be seated at a curved console with a swivel chair. The officer will have the ability to observe property locations and report suspicious events. Modern CCTV systems have a number of options. Most systems allow the viewers to zoom, pan, or tilt the surveillance camera to get a closer or more diversified view.

The advantage of a CCTV system is that it reduces the need for additional personnel. Some record events as they are taking place; others operate on a real-time basis and are monitored by an officer. Other systems are in a recording mode only, which means that they only record events and there is no active monitoring by protection personnel.

CCTV systems can be programmed to begin recording when an alarm is activated. There are a number of different ways in which CCTV systems can operate; however, a CCTV system is only as good as the person monitoring the system. The importance of a well-trained, alert protection officer cannot be overstated. The following guidelines are suggested for monitoring a CCTV system:

FIGURE 6-6

1. Always make sure the system is functioning properly. Report any malfunctions immediately. Test the system. Are the camera angles correct? How are the lighting conditions? Are the cameras properly focused?
2. It is recommended that an officer not be assigned to more than ten TV monitors in a given period of time.
3. Focusing on TV monitors over a period of time can be monotonous. Arrangement should be made to take breaks at least every two hours.
4. Unless circumstances require it, do not monitor one screen too long. Attempt to scan the monitors looking for the unusual.
5. Never leave the monitors unattended! If the job requires that the cameras be actively monitored, do so!
6. Any evidence of unlawful intrusions, suspicious activities, or safety hazards must be reported immediately. When in doubt, report.
7. Never read, socialize, watch television, or listen to a radio while monitoring the system. A protection officer's full attention is required.

Lighting. The primary purpose of protective lighting is to deter crime and accidents, and to provide a sense of well being for employees and visitors. Research has indicated that increased lighting helps to reduce crimes against persons and property and is more cost effective than having more personnel (Frisbie, 1977). Proper lighting serves as a psychological deterrent as well as a means of illuminating an area or object. There are four basic types of protective lighting fixtures:

1. *Floodlights*. These provide a beam of light; effective for illuminating property boundaries or storage areas.
2. *Street Lights*. These lights are used in parking areas and produce a low-intensity, diffused light.
3. *Fresnal units*. These provide a long, narrow beam of light used in perimeter areas such as parking lots or open areas.

FIGURE 6–7

4. *Searchlights.* These lights produce a highly focused beam of light and can illuminate a moving object. Search lights can be mounted on vehicles or at various posts.

There are also several sources of artificial light. The most commonly used are

1. *Incandescent lamps.* These are most often used in fresnal units, street lamps, and floodlight-type lighting. These lights come in many sizes and wattages.
2. *Gaseous-discharge lamps.* These are referred to as *mercury* or *sodium vapor* lights. They provide a more efficient lighting, but are slow to relight when hot and slow to light when cold. The sodium-vapor type often produces an orange glow. These lights are used in parking lots and other perimeter areas.
3. *Quartz lamps.* These are very high voltage lamps and produce an intense white light. These lamps are used in perimeter areas requiring bright light.

Regardless of the type of lighting used, it is important to regularly inspect lighting systems to ensure proper functioning. Report any lights which have malfunctions or otherwise appear damaged. Many criminal intrusions and property accidents can be avoided by providing adequate lighting. This is especially important in areas that provide "hiding places" for criminal intruders (parking lots, stairwells, garages). A protection officer should notify the supervisor of any areas in need of additional lighting.

Conclusion   Physical security is an important part of the overall protection plan. It is important for a protection officer to understand and know how to operate security systems. A protection officer should report damage or other malfunctions to management as soon as detected. This responsibility may require making periodic inspections of equipment and barriers. The use of protective lighting, CCTV cameras, alarms, and fencing all contribute to the protection plan.

Discussion Questions

1. Compare the use of key locks with combination lock systems. Which offers more security?
2. Explain the general operation of an alarm system.
3. Identify the seven types of sensor systems used in business protection.
4. Based on the information provided in Table 6-1, what types of alarm systems could be used in the following locations? There may be more than one type for each location.
   a. Banks (business hours only)
   b. A jewelry store after hours
   c. A safe located in a room
   d. An elevator
   e. A storage room for expensive merchandise
   f. A parking garage
5. Identify some advantages and disadvantages of a CCTV system.

References

FRISBIE, D.W., *Crime in Minneapolis, St. Paul.* MN: Governors' Commission on Crime Prevention and Control, 1977.

GALLATI, ROBERT J., *Introduction to Private Security.* Englewood Cliffs, NJ: Prentice-Hall Inc., 1983.

WALSH, TIMOTHY, AND RICHARD HEALY, *Protection of Assets Manual.* Santa Monica, CA: Merritt Publishing, 1989, pp. 4:17–4:19.

T a b l e 6 – 1
Detectors—An Interior Space Protection Review

Device	Application	Operation	False-Alarm Potential	Advantages	Disadvantages
1. Contact Switches	Doors and windows to detect opening	Metallic contact held in place by magnet; removal of which causes spring to open contact, triggering an alarm. Mechanical contact switches operate on similar principle, but without magnet.	Basically stable. Environmental conditions may cause gap between magnet and contact to widen.	Wide variety of applications and types; i.e., plunger recessed, leaf, wide gap, overhead door, mercury, tilt, tamper, reed.	Surface-mounted contacts subject to internal sabotage. Reed switches may "lock-up" if put on seldom-used door.
2. Shock Sensors	Mounted on surfaces subject to forced entry; doors and window frames, walls, safes, cabinets, roofs	Electronically analyzes shocks or vibrations in terms of frequency and intensity.	When sensitivity is improperly adjusted	Adaptable to various construction materials. Processor enables analysis of signal from sensor, ignoring ambient vibrations or shocks.	Relatively expensive.
3. Traps	Duct work, skylights, air conditioning sleeves, above false ceilings	A cord, held under tension, when loosened, tightened, or cut will open circuit.	Stable.	Ideal for protection of unusual points of entry where other types of devices are impractical or ineffective.	When activated or tripped, manual reset is required.
4. Foil	Mounted on glass, laced inside doors	Metallic tape, which carries current, is applied under tension to glass or wood. Forced entry severs tape and opens circuit.	Hostile environment like heavy traffic, temperature, or wind may cause hairline cuts and intermittent opens	Reliable and stable Visible deterrent Low material cost	Easily damaged in a hostile environment. Installation is labor-intensive.
5. Glass Break Sensors (vibration)	Mounts directly on glass	Mechanically responds to the frequency of breaking glass.	When the sensitivity is improperly adjusted	Relatively inexpensive Covers large area of continuous glass Unit does not need power to operate	Potential false alarms from ambient vibrations
6. Audio Discriminators	Glass Protection	Capable of detecting selected frequencies, such as the air-borne sound of breaking glass	When the sensitivity is improperly adjusted	Omni-directional can can protect several hundred square feet. Can protect several windows as well as multi-pane windows.	Many false alarms due to ambient sources.

Device	Operational Theory	Application	Characteristics	Considerations
7. Glass Break Sensors (piezoelectric)	Mounts directly on glass		Electronically detects the intermolecular noises of breaking glass	Covers large area of continuous glass. Fairly stable device. Unit needs power to operate. Relatively expensive when compared to other types of glass protection. Limited sensitivity adjustment potential.
8. Vibration Contacts	Mounted on surfaces subject to forced entry: doors, window frames, walls, safes, and ceilings		Vibration/shock causes two touching pieces of metal to separate upon impact and open the circuit. When the sensitivity is improperly adjusted	Inexpensive unit cost
9. Ultrasonics	Transmits and receives sound waves. Transmitted sound waves, upon striking a moving object, return to the receiver at a different frequency. This frequency difference (or Doppler Shift) trips an alarm.	Volumetric coverage Warehouse Schools Municipal Buildings	Easily contained; will not penetrate most materials, but may be absorbed. Sound waves will bounce off hard surfaces to fill protected area.	Avoid: Areas with air turbulence Noisy areas, phones, machinery, moving signs, displays, drapes, plants, etc. Severe humidity fluctuations Changing configuration of area (moving stock or equipment) without a corresponding sensor adjustment.
10. Microwave	Radio waves transmission and reception; also works on the Doppler Shift principle	Volumetric coverage Large open areas Factories Warehouses	Will penetrate most nonmetallic materials, but will be reflected off metal surfaces. Reflected waves may "leave" protected areas. Can be mounted in false ceilings for covert detection or to protect unit from tampering.	Avoid: Aiming at vibrating or moving metal surfaces Unwanted signal penetration Fluorescent lighting (at least by 3 feet) *System needs extensive walk-testing due to penetration potential.
11. Passive Infrared	Receives only infrared energy or heat. Unit constantly looks for a relative rapid change in temperature between two separate zones. If intruder enters one zone, the relative change in temperature is sensed and an alarm occurs.	Pattern or array of beams. Large open area Hallways or aisles Offices	Infrared energy will not penetrate most materials. Needs a clear line of sight.	Avoid: Hot or cold drafts on unit Pointing at areas where animals may move Aiming at moving signs, plants Hot environment, above 95°F Moving stock, furniture or equipment that may block unit

Device	Operational Theory	Application	Characteristics	Considerations
12. Photo Electric	Transmission of a pulsed infrared light beam is focused on a receiver (photo-conductive cell), causing current to flow. Interruption of this beam stops the current flow and causes an alarm.	Channel protection Main access areas or corridors Long hallways or warehouse aisles Long unobstructed walls or row windows	Needs a clear line of sight Range 300 to 800 feet Use of mirror allows beam to go around corners	Avoid: Moving objects that may cross beam Areas where animals may move Moving stock, furniture, or equipment that may block beam *Alignment critical, use "guard rail" to protect unit.
13. Mats	When pressure is applied to two metal strips (or electrodes) separated by sponge rubber, they touch and short out the system and cause an alarm.	"Spot" protection Any place where intruder is likely to enter and walk: under windows, in front of safes, cabinets, doors and along corridors	Inexpensive protection Covert detection, used under carpets Pressure sensitive	Avoid: Moving furniture or equipment onto mats Heavy traffic area Wet or moisture prone areas

Note: All devices should be mounted on stable, vibration-free surfaces.
Source: Security World, Jan. 1985, pp. 44–45.

Chapter 7

Physical Force and Defensive Measures

Learning Objectives

After studying this chapter, you should be able to explain the following:

Levels of force

Defensive force

Types of chemical agents

The three classes of batons

When handcuffing is recommended

Deadly force

Shooting policy

Use-of-force encounters

Questions to consider before using force

Firearm training

INTRODUCTION

Critical to the role of private protection is the use of physical force. The use of force should be a rare event and thought of only as a last resort when alternative measures such as verbal commands have failed. Situations such as arrests may require the use of physical force. Before making an arrest, the protection officer must be sure that he or she is acting legally. The officer should have observed a crime being committed before attempting to restrain the person until other authorities arrive. A protection officer could also use force when his or her safety or someone else's safety is jeopardized. If the officer is being attacked by an intruder, there is no alternative but to use force. The force used must not be excessive. This means that the officer does not use force to severely injure or punish someone, but uses only the amount of force necessary to accomplish the objective.

The defensive measures discussed in this chapter are offered as a guide to the use of force. Selected techniques are discussed along with some illustrations. However, the measures and tactics discussed are not meant to be the sum total of knowledge on the use of force. It is highly recommended that the protection officer attend classes and stay up to date on use-of-force strategies. Some jurisdictions require that protection officers be certified in such areas as baton and firearm usage. Whatever weaponry the officer is required to carry, he or she should receive proper training in the use of restraint.

LEVELS OF FORCE

The use of force should be thought of as a measured response to a situation. In other words, the amount of force depends upon the actions of the offender. If the offender is submissive and responsive to commands, then no force other than verbal commands is necessary. If an intruder attacks with a deadly weapon such as a knife in an enclosed area, there may be no alternative to using deadly force, or force likely to cause death to the intruder.

Reasonable force can be justified only after all other means of gaining compliance have failed. The following are situations in which reasonable force may be used by protection officers:

1. Making an arrest
2. Preventing the escape of a dangerous person
3. Overcoming resistance or attack.

The use of force in any situation is designed to control an offender, or to react to the offender's actions. Only the amount of force necessary for the situation should be used. Figure 7–1 describes the various levels of force. Note that as the offender's actions become more threatening, the protection officer's actions likewise will become more defensive. If a trespasser is detained and then attempts to run, what would be the appropriate response? Unless the offender could somehow immediately be caught, the protection officer would probably let him go! If the protection officer caught the offender, he or she would probably handcuff the offender and use some type of firm grip control. The more the offender resists, the more control can be used by the protection officer.

Important to the use of force is when to use it. Protection officers will usually not be involved with shootouts, car chases, and barricaded suspects. As a general rule, the protection officer will not have the same quality of training as police officers. Moreover, he or she usually will not be paid as much as a police officer. The point is to avoid any situations requiring the use of physical force. If

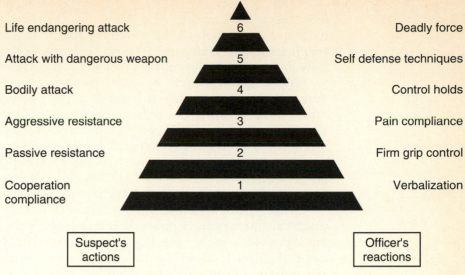

Suspect's actions		Officer's reactions
Life endangering attack	6	Deadly force
Attack with dangerous weapon	5	Self defense techniques
Bodily attack	4	Control holds
Aggressive resistance	3	Pain compliance
Passive resistance	2	Firm grip control
Cooperation compliance	1	Verbalization

FIGURE 7–1   Escalation and de-escalation of force.

the use of force is necessary, it must be reasonable and appropriate. Consider the following questions on when to use force:

1. Are there alternatives to physical force?
2. Is there anyone available to assist me or give me advice?
3. Is the event or crime serious enough to use force?
4. Is it better to let the offender go than get into a physical battle?
5. Is the offender known (former employee)?
6. Do I have the appropriate training?
7. Does my company have a use-of-force policy?
8. How many offenders are involved?
9. Is the offender intoxicated or under the influence of a drug?

These and other questions need to be discussed with the protection officer's supervisor and other protection officers. The protection officer must know what he or she is expected to do. If, for example, the protection officer is expected to stand by at his or her post and simply report a crime that occurs, then that should be clearly understood.

WEAPONS

There are a number of weapons available to the protection officer. Weapons should be regarded as defensive measures only, and should not be carried unless authorized and trained in its usage.

Chemical Agents

There are various types of chemical agents used by law enforcement and protection officers. The three most common are:

1. CN (chloracetuphenone) gas
2. CS (ortho cloro benzalamalononitrile) gas
3. DC (oleo resin capsicum) solution

The *CN gas* is a white crystalline solid resembling granulated sugar or salt. It is classified as a tear-inducing agent. It has an odor resembling grape or locust blossoms; it causes tearing, a temporary feeling of blindness, and a burning or stinging sensation to exposed areas of the skin. An affected person will have difficulty breathing. A heavy concentration of CN in an unvented room could be fatal.

The *CS gas* is a white crystalline powder resembling talcum powder in its pure form. It has a pungent odor and it is a more powerful agent than CN gas. It causes profuse tearing and tightness in the chest. Exposure to CS gas results in an inability to breathe, followed by coughing. It can also produce a general feeling of helplessness and psychological depression.

The *DC solution* is manufactured from a red-pepper base and is more incapacitating to the subject than CN or CS gas, but does not have the after-effects of CN or CS. It is becoming a popular type of defensive chemical agent.

The type of chemical agent used will depend upon company policy. Chemical agents come in pressurized canisters (Figure 7–2). When the button on top is pushed, a steady stream of gas is emitted. The spray will normally have a range of 15 feet; however best results are usually obtained between 8 and 10 feet. The purpose of using a chemical agent is to temporarily incapacitate a person. These agents are designed to give the protection officer an advantage in controlling a person; however, it should be recognized that some persons may not be affected by chemical agents. These persons are usually on certain mind-altering drugs (PCP) or are violent, mentally ill persons.

The following guidelines are recommended in using a chemical agent:

1. Spray directly in the person's face and away from you.
2. Spray only when under attack.
3. Do not spray near other officers. Consider wind factors.
4. Consider using other tactics before using gas.

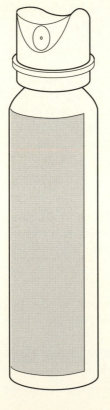

F I G U R E   7 – 2   Chemical irritant.

5. If there are too many people around, withdraw and wait for assistance.
6. Always make sure that the gas is properly stored.
7. Test the canister occasionally to make sure it is functioning.
8. Do not carry chemical agents on your person unless authorized to do so and on duty. It is a crime in some states to carry a chemical agent unless certified.

The effects of chemical agents can last from 5 to 85 minutes. If a person is sprayed, it may be necessary to seek medical treatment. The protection officer should notify his or her supervisor if a person has been sprayed. Remember, policy may require that a report be completed explaining the circumstances under which the protection officer was forced to use a chemical agent. If contaminated with a chemical agent, immediately leave the area, face into the wind, and avoid rubbing the eyes and other irritated areas. Do not flush with water since the pores will close and the burning sensation will remain longer. Shower in luke-warm water and have any contaminated uniforms cleaned. The chemical irritant is another tool which, if used properly, will reduce injuries to protection officers and others. However, it must be considered that the use of a chemical agent depends upon a number of factors. It may not be effective with everyone, and its use should be avoided in a crowd. Other options must be considered before using this defensive measure.

The Baton

The baton is considered an intermediate-range defensive weapon. There are three fundamental classes of batons:

1. The one-handed, short baton (up to 24 inches)
2. The two-handed, thirty-inch baton
3. The two-handed, thirty-six inch baton

The baton has undergone changes throughout the years. The first batons were made of wood (hickory). By the mid-1960s, plastic batons appeared in use. The heavier plastic batons of the 1960s and 1970s hit harder, but were vulnerable to extreme cold temperatures. Currently, the *PR–24 side-handle baton* introduced in 1971 is the most popular police baton (Figure 7–3). A number of protection officers also carry this type. The basic PR–24 baton is 24 inches in length, weighs 24 ounces, and is made of polycarbonite plastic. It has a side handle with a right angle to it about six inches from the end. It is an effective weapon for close-in defensive strikes. It can also be used as a come-a-long. A *come-a-long* is used to remove or direct passively resisting offenders from public disturbance situations. Due to its popularity, usage of this type of baton will be the focus of discussion.

Training for use of the baton varies, but generally requires anywhere from 8–12 hours. The state of California, for example, requires all protection officers

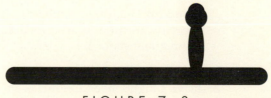

FIGURE 7–3

to receive eight hours of state-mandated training. Anyone caught in possession of a baton without authorization, can be charged with a felony. Training in baton use includes techniques in blocking, counterstriking, restraint, and come-a-long measures. The baton is used to incapacitate rather than inflict serious lifelong injury. There are countless cases where police officers and protection officers have been held civilly and criminally responsible for inappropriate use of a baton.

The baton is considered a defensive and deadly weapon and should not be used unless no other alternative exists. The effectiveness of baton usage is based more on technical proficiency than physical strength. It is imperative that the protection officer receive the required training as well as periodic refresher training. There are a number of training schools available that teach baton usage; law enforcement agencies are also a source of baton training.

Firearms

Most protection officers do not carry firearms. The reason is that most post assignments require observation and reporting skills rather than reliance on physical force. The carrying of a firearm also could lead to serious liability problems. However, those officers who do possess firearms need to recognize the technical, moral, and legal aspects of firearm usage. If required to carry a firearm, a training program is a must. The training should focus not only on the techniques of shooting, but on decision making as well. Knowing when to shoot is just as important as knowing how to shoot. The protection officer should understand the safety features of firearms. When a firearm is not being carried, it must be secured in a safe place. It is the protection officer's responsibility to make sure the firearm is stored safely off duty, handled safely on duty, and always in good working order. As with other types of weapons, a firearm should be considered a last-resort defensive weapon only! Protection officers who patrol alone in high-risk environments should be prepared to respond reasonably to personal threats.

Before undertaking an assignment requiring the possession of a firearm, the protection officer should have received a course of instruction in firearms usage. While some states mandate different training requirements, it has been recommended that all private protection personnel receive specified training and requalification in firearms. The following recommendation is taken from the 1973 Private Security Standards and Goals Task Force Report.

Standards and Goals—Firearms and Uniforms—Task Force Report

Standard 2.6. All armed private security personnel, including those presently employed and part-time personnel should

1. Be required to successfully complete a 24-hour firearms course that includes legal and policy requirements—or submit evidence of competence and proficiency—prior to assignment to a job that requires a firearm; and
2. Be required to requalify at least once every 12 months with a firearm(s) they carry while performing private security duties (the requalification phase should cover legal and policy requirements).

Standard 11.3. Every applicant who seeks registration to perform a specific security function in an armed capacity should meet the following minimum qualifications:

1. Be at least 18 years of age;
2. Have a high school diploma or pass an equivalent written examination;

3. Be mentally competent and capable of performing in an armed capacity;
4. Be morally responsible in the judgment of the regulatory board;
5. Have no felony convictions involving the use of a weapon;
6. Have no felony or misdemeanor convictions that reflect upon the applicant's ability to perform a security function in an armed capacity;
7. Have no physical defects that would hinder job performance; and
8. Have successfully completed the training requirements for armed personnel set forth in Standards 2.5 and 2.6.

The protection officer must understand the shooting policy of his/her agency. This means know the law of self defense, since law and policy are related. In policing, law enforcement officers are generally trained to shoot when attacked with a deadly weapon, when another is under deadly attack, or to stop a dangerous fleeing felon such as a robbery suspect who has shot someone. It is hoped that the protection officer will never have to fire his or her weapon. However, no assignment can be considered "safe," nor can any time, day, month, or location. A protection officer can be killed in daylight, at night, or on weekends. Wearing a uniform represents authority. To some offenders, the protection officer also represents a threat. This may be especially true if the protection officer is assigned to a bank, industrial complex, armored-car service, or to some housing projects.

PATTERNS OF ENCOUNTERS

Research on police shootings have indicated some general patterns. While these patterns may not be relevant to private protection, it is worth discussing some of these patterns of encounters. This discussion may help the protection officer consider the consequences of being caught unprepared (Adams and others, 1980).

Distance

In almost 85 percent of actual shootouts between officers and suspects, there is a span of less than 7 yards (most shooting ranges involve targets set at 7 to 50 yards). Also, in the majority of cases, officers are killed less than 10 feet from their assailants (in half of the cases, only 5 feet or less)—often close enough to touch each other. In other words, the offender that shoots at you is likely to be less than half the length of the vehicle, or the length of your arm away from you.

Light

Most fatal officer shootings occur at night or in areas where the light is extremely dim (hallways, basements).

Time

The protection officer will only have 2 to 3 seconds to react or respond in the majority of confrontations. These are sudden activities and there is almost never enough time to cock a gun, stand in profile, and take careful aim at the target. Therefore, the speed with which the protection officer reacts to danger is more critical than the amount of available ammunition.

Location

Outdoor shootings outnumber those that take place inside by 2-1/2 to 1. For the most part, these shootings occur on streets and sidewalks or in alleys and yards. That means the protection officer must be acutely aware of the immediate vicinity (motorists, people in homes, children playing, fellow officers, hostages). Take into account bystanders' welfare because the protection officer will be held legally responsible for any stray shots.

Assailants

Unlike the shooting range where stationary targets are shot one at a time, at least four out of ten gunfights involve more than one assailant. Criminals employ backup men who are not easily identifiable as offenders and may not be noticed until they open fire.

Weapons

About 15 percent of street suspects will shoot with a shotgun or rifle. Also, their guns may be ballistically superior. Seven out of 10 officers killed die from handgun wounds. More than 80 percent of guns used against officers are the same caliber or smaller than the officer's. Many of these guns have been stolen or bought from fences (people who deal in stolen goods). A variety of high-quality weapons are available to criminals from illegal sources. Remember, an intruder is likely to be armed, to fire first, and to have a gun that is ballistically superior to yours.

Persons who Shoot

The odds are better than 40–60 that once a protection officer confronts a suspect who has a gun, the suspect will shoot at the protection officer. This act may be done out of panic, desperation, confusion, anger, fear, derangement, exasperation, revenge, intoxication, hallucination, political zeal, or suicidal yearning. Whatever the motivation, a sizeable proportion of suspects with guns are not intimidated by a uniform. On the contrary, it may well be the uniform and the authority it represents that incites them to shoot. Statistics have given us patterns to look for:

- Of those who kill officers, more than 96 percent are male and nearly half are black; most are less than 30 years of age.
- The most likely to confront law enforcement officers while in the company of others are under 21 and black; those over 30 and white will generally act alone.
- These assailants are pros; over 60 percent have prior criminal records with about 40 percent for violent crimes (murder, rape, robbery, or assault).
- A large number are on probation or parole. Although most are not trained marksmen, they are known to practice a great deal with firearms.
- Many are knowledgeable in police tactics and have developed plans for confronting the police.
- These assailants are frequently sociopaths; they will shoot without remorse or conscience. Do not expect rationality or compassion.

Other Attitudes

While the assailant's profile prompts him to shoot, the protection officer copes with a number of constraints that conspire to keep him or her from shooting. A protection officer will weigh a number of concerns before shooting: department policy on the use of deadly force; and the welfare of his or her family, partner, or bystanders. Protection officers often opt not to shoot because of strong moral objections to killing. Overconfidence is often the greatest roadblock in shoot or don't-shoot circumstances. Do not make the mistake of thinking yourself invulnerable. This attitude can lead you to underestimate your adversary.

Approaching Danger

Surviving the job requires you to:

- Recognize that your motivations to shoot are dramatically different from those of the suspect.
- Anticipate the legal, moral, and psychological implications of a shooting; resolve these before the confrontation.
- Be prepared to take a life in order to protect your own or someone else's life.
- Make your learning and practicing with firearms as realistic as possible.
- Maintain preparedness through planning and proper physical fitness.

The general rule is to not draw your weapon unless you are to shoot. This rule would most likely apply when

1. You are approaching a situation where you know or suspect someone has a weapon.
2. You are involved in a building search where a possible intruder may be present.
3. You have reason to fear for your safety or the safety of others.

The question arises as to what to do if an intruder is discovered but flees from the scene. For example, if a burglary in progress is discovered, what is the proper action to take? Remember that the use of deadly force is a last resort. A protection officer who shoots a fleeing unarmed intruder will undergo intense questioning. The consequences of such an action could include criminal and civil actions, job loss, and living with the memories of having killed someone. It is better to let someone go than to shoot. Unless the protection officer has no other choice (self defense), it is better to retreat and get a description of the offender than to use deadly force. The firing of warning shots is not recommended because there are too many serious consequences that can result from stray bullets.

When approaching an intruder or one who may pose a threat to your safety, do not encourage or otherwise force a confrontation. In other words, remain a safe distance from the person. Do not allow the person to get in a position where he can attack you or grab your weapon. If your weapon is drawn, remain a safe distance from the offender.

HANDCUFFING

The primary purpose in using steel handcuffs and other restraints is to maintain control of an arrestee, provide safety for both the officer and arrestee, and minimize the possibility of the situation escalating. Handcuff use generally is not subject to rigid guidelines. The protection officer should decide whether to use handcuffs based on the circumstances of each particular incident.

As a protection officer you may have to handcuff a suspect. Use common sense and sound judgment when making this decision. Some of the factors involved include, but are not limited to, the following:

- Possibility of the detainee escaping
- Possibility of the incident escalating
- Potential threat to officers and other persons
- Knowledge of the detainee's previous encounters with law enforcement

In law enforcement, felony arrestees are normally handcuffed; however, there may be extenuating circumstances indicating that handcuffing would be inappropriate. In cases of misdemeanors, handcuffing is often discretionary. Handcuffs are normally used if there is a belief that the arrestee may attempt to escape or if the safety of the officer is at stake.

Unless the specific situation warrants otherwise, an arrestee should be handcuffed behind the back. It is important to follow some basic procedures when applying handcuffs: 1) handcuff the offender's hands behind the back with the hands placed back-to-back (see Figure 7–4), 2) avoid injury to the wrists when applying the handcuffs, 3) check the handcuffs to be sure they are neither too tight nor too loose, and 4) double-lock the handcuffs.

It is important to constantly monitor a detainee because handcuffs are temporary restraints and not escapeproof. Be alert for an attempted escape and continue to check the suspect while in transit. It is a good idea to practice handcuffing periodically with other officers.

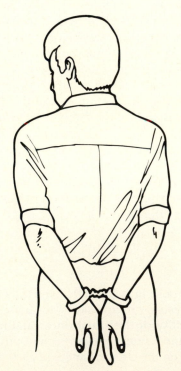

FIGURE 7–4

Conclusion   The use of physical force is considered a last resort in attempting to control another person. Before using any force, be sure that alternative actions have been exhausted. If time permits, contact a supervisor or other officer before using force. If force is necessary, use only the amount of force necessary to control or otherwise stop an offender. Report the degree of force used and the reasons for its use. If there is injury to another, seek medical assistance immediately even if the injury appears to be slight. If possible, use verbal commands rather than physical force in order to control another person. Remember, the use of any weapon or physical-control procedure requires training and retraining. Do not accept any assignment requiring the use of weapons unless proper training and instruction has been provided.

Discussion Questions

1. What factors determine the level of force to be used by a protection officer?
2. What level of force, if any, would be acceptable in the following situations? Assume the officer is armed with a firearm, baton, and a chemical agent.
   a. An unarmed shoplifter flees a store and refuses to stop.
   b. An officer observes a burglar fleeing a warehouse refusing to stop.
   c. A mentally disturbed husband armed with a shotgun holds his former wife and her co-workers hostage.
   d. An unarmed robbery occurs in front of a crowded department store.
   e. A purse snatch is witnessed by a protection officer.
3. Describe a situation which would justify the use of deadly force.
4. What firearm training does the Private Security Standards and Goals Task Force Report recommend for security officers?

References

GALLATI, ROBERT R.J., *Introduction to Private Security*. Englewood Cliffs, NJ: Prentice-Hall, Inc., 1983, p. 272.

Task Force Report on Private Security Standards and Goals. U.S. Government Printing Office Washington, D.C., 1973.

ADAMS, RONALD J., THOMAS M. MCTERNAN, AND CHARLES REMSBERG, *Street Survival Tactics For Armed Encounters*. Northbrook, IL: Calibre Press, 1980.

The Protection Officer's Role in Health and Safety

Learning Objectives   After studying this chapter, you should be able to explain the following:

Disaster response

Fire prevention

First aid and CPR

Handling bomb threats

Hazardous materials

Fire extinguishing

Occupational Safety and Health Act (OSHA)

Work-place hazards

Classes of fire

Americans With Disabilities Act (ADA)

INTRODUCTION

As a protection officer, one of your concerns will be safety and health maintenance. Safety in the workplace requires awareness of hazards that may cause injury to employees or invitees. Safety maintenance refers to fire prevention, hazardous materials recognition, and emergency response to unusual disasters. This chapter describes recommended principles and procedures in health and safety, including first aid and cardiopulmonary resuscitation (CPR).

LAWS REGULATING HEALTH AND SAFETY

There are a number of federal agencies that regulate health and safety standards (Table 8–1).

TABLE 8–1
Federal Agencies Regulating Health or Safety

Consumer Product Safety Commission (CPSC)   Establishes mandatory safety standards governing the design, construction, contents, performances, and labeling of hundreds of consumer products.

Environmental Protection Agency (EPA)   Regulates air, water, and noise pollution; regulates waste disposal and specific chemicals considered hazardous to people and the environment; registers pesticides and regulates their use; administers cleanup of hazardous dumps; monitors other potential pollutants.

Nuclear Regulatory Commission (NRC)   Licenses the construction and operation of nuclear reactors and other facilities; licenses the possession, use, transportation, handling, and disposal of nuclear materials; develops and implements rules and regulations governing licensed nuclear activities; licenses the export and import of uranium and plutonium.

Occupational Safety and Health Administration (OSHA)   Develops and enforces mandatory job safety and health standards; maintains reporting and record-keeping procedures to monitor job-related illnesses and injuries.

Food Safety and Quality Service   Regulates the meat, poultry, and egg industries for safety and purity by inspecting all meat, poultry, and eggs shipped in interstate and foreign commerce.

Food and Drug Administration (FDA)   Regulates the purity and labeling of food, drugs, cosmetics, and medical devices to protect the public against potential health hazards from these products.

Mine Safety and Health Administration   Develops and promulgates mandatory safety and health standards, ensures compliance with such standards, and proposes penalties for violating standards.

Federal Aviation Administration (FAA)   Establishes and enforces rules and regulations for safety standards covering all aspects of civil aviation.

Materials Transportation Bureau   Develops and enforces equipment and operating safety regulations for the transportation of all materials by pipeline; designates substances as hazardous materials and regulates their transportation in interstate commerce.

National Highway Traffic Safety Administration (NHTSA)   Develops mandatory minimum safety standards for domestic and foreign vehicles sold in the United States; develops safety and wear standards for tires.[1] All States have enacted *worker compensation laws*. Worker compensation provides for lost wages, rehabilitation, medical costs, and death benefits for dependents of employees killed on the job.

[1]Organizational Security, *Health and Safety*, p. 29

Source: Adapted from Regulation: Process and Politics. *Congressional Quarterly*, 1982, 137–143.

In 1970, Congress passed the Occupational Safety and Health Act (OSHA). The purpose of OSHA is to ensure that every employer engaged in a business affecting commerce will provide employees with safe working environments. This act creates occupational health and safety standards which are binding upon all employers subject to the act's jurisdiction;. Publication and enforcement of OSHA standards is the responsibility of the U. S. Department of Labor. Businesses affected by OSHA should have a copy of the standards. Copies can be found in most libraries, or ordered from the U. S. Government Printing Office, Washington, D. C.

If OSHA standards are not met by the employer, the employer could be subject to a fine, criminal penalties, or possible closures. Inspections are conducted when health and safety violations occur, employee complaints are received, or when occupational accidents occur. However, the United States Supreme Court has ruled that warrantless inspections may not occur without an employer's consent. Inspections of a business may occur without an owner's consent if there is reason to believe the business is violating health and safety standards. In these cases an authorized judicial consent is needed. Some of the safety policies and procedures required by OSHA as well as insurance companies include conforming to safe procedure while operating equipment; and the wearing of protective glasses, shoes, and other clothing. There are a number of federal agencies that regulate health and safety standards in the workplace. A protection officer may be required to make inspections and report violations.

Work-Place Hazards. In 1987, the National Safety Council reported that approximately 11,000 accidental deaths and nearly 2 million serious injuries occurred in the work place. A number of these accidents may have been avoided through the recognition of potential safety hazards. In the event an employee is injured on the job, states have enacted *worker compensation laws*. Worker compensation provides for lost wages, rehabilitation, medical costs, and death benefits for dependents of employees killed or injured on the job. It is unfortunate, however, that a few unhappy employees abuse these protections by fraudulently sustaining injuries on the job in order to claim benefits. As a protection officer, it may be your responsibility to not only detect, but to investigate work-place injuries.

There are a number of work-place hazards. The following list describes some common safety hazards that can cause injuries. A protection officer must look for and report the following:

Common Safety Hazards

1. *Floors, aisles, stairs, and walkways.*

    Oil spills or other slippery substances which might result in an injury-producing fall.

    Litter, obscured hazards such as electrical floor plugs, or combustible materials.

    Electrical wire, cable, pipes: crossing aisles which are not clearly marked or properly covered.

    Stairways which are too steep or have no nonskid floor covering; railings which are inadequate or nonexistent, or in a poor state of repair.

    Overhead walkways which have inadequate railings, are not covered with nonskid material, or which are in a poor state of repair.

    Walks and aisles which have not been cleared of snow or ice, are slippery when wet, or in a poor state of repair.

2. *Doors and emergency exits.*

Doors that are ill fitting, stick, and which might cause a slowdown during emergency evacuation.

Panic-type hardware which is inoperative or in a poor state of repair.

Doors which have been designated for emergency exit, but are locked, chained, or not equipped with panic-type hardware.

Doors which have been designated for emergency exit, but are blocked by equipment or debris.

Missing or burned-out emergency exit lights.

Nonexistent or poorly marked routes leading to emergency exit doors.

No evacuation plan.

3. *Flammable and other dangerous materials*

Flammable gases and liquids which are uncontrolled, in areas in which they might constitute a serious threat.

Radioactive material not properly stored or handled.

Paint or painting areas which are not properly secured or ventilated. Gasoline pumping areas located dangerously close to flames or spark-producing operations.

4. *Protective equipment or clothing.*

Workers in areas where toxic fumes are present who are not equipped with nor using respiratory protective apparatus.

Workers involved in welding, drilling, sawing and other eye-endangering occupations who have not been provided or are not wearing protective eye covering.

Workers in area requiring the wearing of protective clothing due to exposure to radiation or toxic chemicals who are not using such protection.

Workers engaged in the movement of heavy equipment or materials who are not wearing protective footwear.

Workers who require prescription eyeglasses who are not provided or are not wearing safety lenses.

5. *Vehicle operation and parking.*

Forklifts which are not equipped with audible and visual warning devices when backing.

Trucks which are not provided with a guide when backing into a dock, or not properly chocked when parked.

Speed violations by cars, trucks, lifts, and other vehicles being operated within the protected area.

Vehicles which are operated with broken, insufficient, or nonexistent lights during the hours of darkness.

Vehicles which constitute a hazard due to poor maintenance procedures on brakes and other safety-related equipment.

Drivers not licensed to operate type of vehicle; unauthorized use; unsafe acts.

6. *Machinery maintenance and operation.*

Frayed electrical wiring which might result in a short circuit or malfunction of the equipment.

Workers who operate presses, work near or on belts, conveyors, and other moving equipment who are wearing loose-fitting clothing which might get caught and drag them into the equipment.

Presses and other dangerous machinery not equipped with the required hand guards, automatic shut-off devices, or dead-man controls.

7. *Welding and other spark-producing equipment.*

Welding torches and spark-producing equipment being used near flammable liquid or gas storage areas.

The use of flame-producing or spark-producing equipment near wood shavings, oily machinery, or where they might damage electrical wiring.

8. *Miscellaneous Hazards.*

Medical and first-aid supplies not properly stored, marked, or maintained.

Incorrect or incomplete color coding of hazardous areas or materials.

Broken or unsafe equipment and machinery not properly tagged with a warning of its condition.

Electrical boxes and wiring not properly inspected or maintained.

Emergency evacuation routes and staging areas not properly marked or identified (Fischer and Green, 1992).

It is important for the protection officer to know and understand the company rules and policies regarding safety procedures. The checklist in Figure 8–1 is recommended for use in detecting potential safety problems. Checking for hazards should be part of a protection officer's daily patrol duties. A more detailed check should be done every month. In the event a hazard is encountered, report it to your supervisor immediately. Remember, a protection officer's primary role is to observe and report any unusual hazard or incident. Awareness of work place hazards or violations may save lives and reduce injuries.

The Law and the Americans With Disabilities Act (ADA)

In 1990, President Bush signed the first comprehensive civil rights law for people with disabilities. The ADA took effect in 1992 prohibiting private employers, state and local governments, employment agencies, and labor unions from discriminating against qualified individuals with disabilities. Persons with disabilities include those with physical or mental impairments that limit their life activities. The act protects former drug users who have been rehabilitated.

An employer is not required to lower job standards to accommodate the disabled or provide personal-use items (for example, hearing aids). An applicant for a protection position requiring extensive foot patrol cannot have certain physical limitations. However, there may be assignments such as dispatching or administrative tasks that could be performed by a disabled person.

The ADA is important in the protection business because protection officers may be required to inspect a premises to ensure that certain services and accommodations are available to the disabled. Violations of the ADA are enforced by the federal Department of Justice. Violations can result in substantial fines against a business.

In summary, the ADA is designed to protect the disabled from discrimination in seeking jobs, promotions, or assignments they could perform. A protection officer's employer must understand this act and take steps to implement programs and policies.

FIRE PREVENTION AND RESPONSE

A primary safety concern for the protection officer is fire prevention. Fire prevention is a serious business. Consider the following as evidence of fire threats in the hotel industry.

Monthly Safety Check				
General Area		Department _____ Date _____		
Floors Condition		Supervisor _____		
Special Purpose Flooring		Indicate Discrepancy using a checkmark		
Aisle, Clearance/Markings		First Aid		
Floor Openings, Require Safeguards		First Aid Kits		
Railings, Stairs Tem./Perm.		Stretchers, Fire Blankets, Oxygen		
Dock Board (Bridge Plates)		Fire Protection		
Piping (Water–Steam–Air)		Fire Hoses Hung Properly		
Wall Damage		Extinguisher Charged/Hung Properly		
Ventilation		Access to Fire Equipment		
Other		Exit Lights/Doors/Signs		
Illumination—Wiring		Other		
Unnecessary/Improper Use		Security		
Lights on During Shutdown		Doors/Windows Etc. Secured When Required		
Frayed/Defective Wiring		Alarm Operation		
Overloading Circuits		Department Shutdown Security		
Machinery Not Grounded		Equipment Secured		
Hazardous Location		Unauthorized Personnel		
Housekeeping		Other		
Floors		Machinery		
Machines		Unattended Machines Operating		
Break Areas/Latrines		Emergency Stops Not Operational		
Vending Machines/Food Protection Rodent, Insect, Vermin Control		Platforms/Ladders/Catwalks Instructions to Operate/Stop Posted		
Waste Disposal		Maint. Being Performed on Machines in Operation		
Vehicles		Guards in Place		
Unauthorized Use		Pinch Points		
Operating Defective Vehicle		Materials Storage		
Reckless/Speeding Operation		Hazardous & Flammable Material Not Stored Properly		
Failure to Obey Traffic Rules		Improper Stacking/Loading/Securing		
Other		Improper Lighting, Warning Signs, Ventilation		
Tools		Other		
Power Tool Wiring				
Condition of Hand Tools				
Safe Storage				
Other				

FIGURE 8–1

- November 21, 1980—MGM Grand Hotel, Las Vegas, NV (84 dead)
- December 4, 1980—Stouffer's Inn, Harrison, NY (26 dead)
- January 17, 1981—Inn of the Park, Toronto, Canada (6 dead)
- February 10, 1981—Las Vegas Hilton, Las Vegas, NV (8 dead)

In 82 days, 168 deaths occurred in hotel fires alone (Coakley, 1982). The major cause of death in building fire is caused by toxic gas followed by smoke, heat (high temperature), carbon dioxide, fear, and panic. Flames are rarely the primary cause of death. Most fires are caused by carelessness or ignorance. An effectively trained protection officer and ongoing fire safety program will reduce the potential loss due to fire.

An officer's duties may include observing and reporting hazardous fire conditions and making sure emergency equipment is operable. When reporting for duty, test the fire sensors and alarm systems to be sure they are working. A list of the telephone numbers for the fire department, police department, and ambulance service should be available. Security posts must have emergency telephone numbers listed. While on patrol, fire prevention is one of your most important duties; be alert for fire hazards such as oily rags, open solvent containers, leaking storage drums, and waste-paper clutter. Defective electrical wiring and equipment left running also present fire hazards. Be aware and have a plan.

Fire Extinguishers

It is important to know the location of all the extinguishers on your patrol and be familiar with the different types of firefighting equipment. There are three basic elements of a fire: 1) air (oxygen), 2) heat, and 3) fuel. The potential for fire exists when these elements are present; however, in order to have a fire all three elements must be present. As indicated in Figure 8–2, a *fire* is a chemical reaction called *combustion* in which fuel and oxygen combine to give off large quantities of heat. By removing one of these elements, a fire cannot exist. A fire can be put out by 1) smothering (removing oxygen), 2) isolating (removing fuel), 3) cooling (removing heat). However, only approved fire extinguishers can be used on certain fires. Fire extinguishers are labeled according to the type of fire in which they can be used. The local fire department is a good training source for extinguisher use.

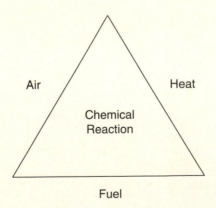

Air—Smother by Diluting or Removing Air
Heat—Reduce by Lowering Temperature
Fuel—Remove Supply or Source

FIGURE 8–2   The fire triangle.

There are generally four classifications of fires requiring specific extinguishers:

- *Type A Fire* Extinguishers are used for ordinary combustible fires such as trash, paper, cloth, rubber, many plastics, and wood. This is a water-based extinguisher.
- *Type B Fire* Extinguishers are used for flammable liquids such as gasoline, fuels, paint, oils, kitchen grease, solvents, and tar. This type of extinguisher has a smothering effect.
- *Type C Fire* Extinguishers contain nonconducting extinguishing agents and are used on fires caused by electrical equipment such as monitors, motors, switchboards, and faulty wiring. In the event of an electrical fire, turn off the power as quickly as possible. This is a multipurpose extinguisher for use on A, B, and C types of fires.
- *Type D Fire* Extinguishers are used for fires that are caused by combustible metals. Carbon-dioxide extinguishers are used for vehicle fires. Patrol vehicles should have one mounted inside for easy accessibility (Cote and Linville 1986).

Types of Fire Extinguishers

As discussed, there are a number of extinguishers available for fire fighting, depending upon the class of fire.

The following is a list of extinguishers commonly found in industrial settings:

1. *Soda Acid.* This extinguisher contains water, bicarbonate of soda, and a small bottle of acid. To use, simply turn it upside down. It is considered to be the most common type of extinguisher. Use this extinguisher in a Class A fire. Never use it on electrical fires unless the electricity has been turned off. This type of extinguisher will freeze in cold buildings; therefore it must be kept in proper storage.

2. *Antifreeze.* This extinguisher works the same way as a soda acid extinguisher, however, it will not freeze. It is also used for Class A fires and, as with the soda-acid extinguishers, it should not be used on electrical fires.

3. *Foam Extinguishers.* This type uses foam to put out fires. The extinguisher blankets the fire with foam which keeps the oxygen away and subsequently the fire goes out. Use this type for Class B fires—flammable liquids. Foam extinguishers can also be used for Class A fires but should not be used on Class C or D fires.

4. *Dry Chemical.* These extinguishers form a cloud of chemicals which keeps oxygen away from the fire and therefore the fire goes out. They can be used for all classes of fires depending on the chemicals in each. The label clearly states which kinds of fires they will put out. These extinguishers are useful for electrical fires and for Class B fires.

5. *Carbon Dioxide.* These are very useful for firefighting because carbon dioxide is heavier than air, it will not burn, it is very cold, and it will not conduct electricity. Use on Class B fires. Also, because it will not conduct electricity, this extinguisher is good for Class C fires.

6. *Water.* This is the oldest type of fire extinguisher and you may have water in a water pump extinguisher or just a bucket to fill. Therefore, if nothing else is available, use it. Good for *Class A* fires. *DO NOT USE for Class C fires* (because water conducts electricity) *or for Class B* fires, flammable liquids.

7. *Halogenated Agents (Halon).* These agents put out fires by obstructing the chemical reaction between fuel and oxygen. There are two basic halogenat-

ed agents used: Halon 1211 (used in portable extinguishers), and Halon 1301 (a halon suppression system). These are generally used in high-tech circumstances to safeguard electrical components or valuable materials (libraries, archives).

Procedures for Fire Prevention

Make sure that fire extinguishers are in good working order and that the date on the inspection seal is current. If there is a gauge that indicates the extinguisher's charge, check it. Sometimes they lose their charge before inspection time. You should also inspect the fixtures and hoses in wall cabinets regularly. Knowing their location and the location of fire call boxes could save lives. Find out whether alarms only ring internally or are connected directly to the fire department.

Fire exits must be clearly marked and kept clear; exit lights (bulbs) must be checked. Be sure and report any blocked exits so management may correct them. Automatic fire doors should also be kept clear. Check all crash-bar doors to be sure they open properly and the exit alarms sound. You need to know the location of the sprinkler shut-off valve controlling the overhead sprinkler system.

Procedures for Responding to a Fire

If you discover a fire, and the alarm is not connected directly to the fire department, your first priority is to call the fire department. Be prepared to direct firefighters to the location of the fire. If another officer is available, that officer should try to control the fire with emergency equipment. If confronted with a small fire, make sure you use the proper extinguisher. Stay near an exit so you will have an immediate escape route. Aim the extinguisher at the base of the fire and then move the spray forward. If the fire is in a small room, stay outside and shoot the stream in. If the fire gets out of control, get out, and close the door to prevent the fire from spreading. Most businesses conduct fire-evacuation drills (necessary for insurance purposes) so employees will know what to do if there is an emergency. In a real emergency, a protection officer's calm, cool manner can help prevent panic and hysterical actions.

The telephone in the elevator is an important part of the emergency reporting system. However, it is important to stay out of elevators during fires. Make sure it is working properly. Telephones should be checked on each tour of duty. In short, have a plan and know what to do in case a fire occurs.

BOMB THREATS

Some businesses are particularly vulnerable to the possibility of bombs being planted by terrorists, arsonists, or former employees seeking revenge on management. Although most bomb threats are false, every threat should be taken seriously. In other words, bomb threats should be considered real unless proven otherwise. Protection officers should be aware of bomb delivery and placement. Bombs can be placed by employees, visitors or delivery persons. They can be carried in purses, briefcases, lunch boxes, vendors cars, express mail, parcel post, and trash containers. There are numerous ways in which a determined person could place a bomb within a building.

In the event you observe a package, container, or device that resembles something suspicious, contact your supervisor immediately. Suspicious objects may be found in restrooms, closets, stairways, elevators, telephone terminal closets, and receiving platforms. If there is a bomb threat, there should be a security policy on how to respond. As a protection officer, your responsibilities would most likely include searching for the object. Unless otherwise directed by the supervisor, do not discuss the bomb threat with other employees or visitors. The decision to evacuate a building is the responsibility of management or the fire department. If the structure is evacuated, the protection officer must ensure that all persons are removed from the premises. During the course of a search, do not touch or otherwise disturb anything that may resemble a bomb. Remember, the role of security in bomb searches is to identify and isolate suspicious items. The removal of such items is the responsibility of bomb-disposal units or other responsible law enforcement personnel. The following guidelines are suggested for security personnel:

- Do not touch anything that looks unusual or suspicious.
- Notify your supervisor immediately.
- Keep all unauthorized persons away and seal or block the area.
- If an explosive device is found in a room during a search, property damage can be minimized by opening windows and doors.
- Do not use radios since some bombs can be activated by radio waves.
- Remain calm.

In the event that a protection officer receives a bomb threat over the phone, certain guidelines should be followed. The checklist taken from the FBI Bomb Data Center (as shown on page 126) should assist the protection officer in recording needed information. In short, gather as much information as possible about the threat. Try to keep the person talking because the information will be used to assist management and law enforcement personnel in determining the nature of the threat.

If a bomb threat is received, keep the caller on the line and ask for information. Record the conversation, make note of the caller's voice, listen for background noises, and remain calm. The smallest detail may help experts determine whether the threat is real or a hoax. As soon as all available information has been gathered, call the security director and notify your supervisor. The local police and the Federal Bureau of Investigation (FBI) should also be notified.

The Bomb Search

If a bomb search is conducted, it can be done by a protection officer, a team, or the building occupants. An occupant search means that each individual looks around his immediate work area for any strange or foreign objects. This must be done by the employees since the search team will not know what belongs and what doesn't. If a strange object is found, don't touch or move it. Evacuate the immediate area, prevent further entry, and report. Regardless of what type of search is used, do not immediately institute evacuation procedures. Many bombers will only have access to public areas and evacuation may take you into harm's way. If evacuation is necessary, the bomb search team will make sure the route is clear. Have employees take personal belongings with them. This will save the search team time and work. During evacuation, walk, don't run. Don't use elevators, don't turn power switches off or on, don't use a radio, or move objects. Most bomb threats are hoaxes made by people seeking attention; however, they all must be treated as real.

There are a number of methods for conducting bomb searches. If a particular room, floor, or area is suspected of containing a bomb, a search may be necessary. Two recommended methods are the strip search and zone search. Both require several personnel and employ a divide-and-conquer approach.

FBI Bomb Data Center

Place This Card Under Your Telephone

QUESTIONS TO ASK:

1. When is bomb going to explode?
2. Where is it right now?
3. What does it look like?
4. What kind of bomb is it?
5. What will cause it to explode?
6. Did you place the bomb?
7. Why?
8. What is your address?
9. What is your name?

EXACT WORDING OF THE THREAT:

_____
_____
_____
_____
_____
_____
_____

Sex of caller: _____

Race: _____

Age: _____

Length of call: _____

Number at which call is received:

_____

Time: _____

Date: _____

REPORT CALL IMMEDIATELY TO:

_____

Phone number _____

_____

Date _____

Name _____

Position _____

Phone number _____

CALLER'S VOICE:

☐ Calm   ☐ Nasal
☐ Angry   ☐ Stutter
☐ Excited   ☐ Lisp
☐ Slow   ☐ Raspy
☐ Rapid   ☐ Deep
☐ Soft   ☐ Ragged
☐ Loud   ☐ Clearing throat
☐ Laughter   ☐ Deep breathing
☐ Crying   ☐ Crackling voice
☐ Normal   ☐ Disguised
☐ Distinct   ☐ Accent
☐ Slurred   ☐ Familiar
☐ Whispered

IF VOICE IS FAMILIAR, WHO DID IT SOUND LIKE?

_____
_____

BACKGROUND SOUNDS:

☐ Street noises   ☐ Factory machinery
☐ Crockery   ☐ Animal noises
☐ Voices   ☐ Clear
☐ PA system   ☐ Static
☐ Music   ☐ Local
☐ House noises   ☐ Long distance
☐ Motor   ☐ Booth
☐ Office machinery   ☐ Other

_____
_____

THREAT LANGUAGE:

☐ Foul   ☐ Incoherent
☐ Irrational   ☐ Taped
☐ Well spoken (educated)   ☐ Message read by
                                          threat maker

REMARKS:

_____
_____

The strip-search pattern shown in Figure 8–3, involves the separation of a series of lanes down which one or more persons advance. The searchers begin at the starting point and advance down their respective lanes, reverse their direction and continue like this until the area has been thoroughly searched. When physical evidence is found all searchers stop until it is properly handled; the search then resumes in the manner described previously. The zone search shown in Figure 8–4 divides an area into four large quadrants. Each quadrant is searched. If the area to be searched is particularly large, then a variation of the zone would be to subdivide the larger quadrants into four smaller quadrants.

To summarize, always carry a checklist which has emergency phone numbers. In almost all emergencies, an officer's primary duty is to notify the proper authorities before taking any action. Compile an emergency chain-of-command call sheet. Know who the emergency coordinator is. In the event of a bomb threat, gather all available information, then notify the proper authorities. Cooperate with the bomb team; assist them with the search and evacuation if necessary.

HAZARDOUS MATERIALS INCIDENTS

A number of businesses use a variety of chemicals. Some of these chemicals are harmful if improperly handled. *Hazardous material* is defined according to the Department of Transportation (DOT) as "any substance or materials in any form or quantity which poses unreasonable risk to safety and health and property when transported in commerce." In subdividing hazardous materials, the Department of Transportation uses two subclassifications:

1. *Hazardous substance* Any material, which when discharged into or upon the navigable waters of the United States or joined shorelines, may be harmful to the public health or welfare of the United States including but not limited to fish, shellfish, wildlife, and public or private shorelines or beaches.
2. *Hazardous wastes.* Any material that may impose unreasonable risk to health, safety, or property when transporting for the purpose of treatment, storage, or disposal as waste.

Hazardous materials can cause serious, even life-threatening problems and must be dealt with immediately. Handling highly toxic liquids, solids, or gases or other highly corrosive materials requires extreme caution. Depending upon the work environment, protection officers may encounter hazardous materials in the course of their duties.

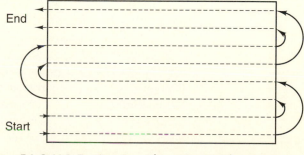

FIGURE 8–3   The strip-search pattern.

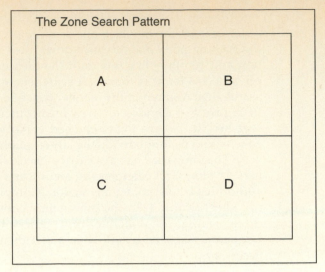

The Zone Search Pattern

| A | B |
| C | D |

FIGURE 8-4

A hazardous materials incident can cause hysteria, a lack of information, and poor visibility. It is often the protection officer who will most likely detect and respond to such an incident. Protection-officer training can make the difference between a major or minor loss of property, environment, or lives. Protection personnel need to know 1) which hazardous materials are typically on site and how they should be contained or stored safely; 2) the conditions which unleash toxic, flammable or volatile properties; and 3) where emergency supplies are kept. A protection force trained to recognize these conditions is invaluable as the first line of defense against intrusions.

Hazardous Chemicals

Federal standards define *hazardous chemicals* as any chemical with the potential to cause a physical or health hazard (the chemicals commonly found in homes are exempted from this standard). Table 8–2 in this section identifies some of the physical and health hazards.

Improper exposure to chemicals can aggravate existing mental conditions or cause acute or chronic health problems. Temporary or long-term effects can result from a one-time overexposure to a chemical or from repeated small exposures over a long period of time. Chronic effects include damage to bodily organs, cancer, birth defects, and sterility. Symptoms which show up soon after improper exposure to a chemical are: eye irritation, nausea, dizziness, skin rashes, and headaches. Symptoms tell you what you have been exposed to. If you know you have been exposed to a chemical, or if you exhibit any of the symptoms, begin emergency procedures immediately. Get help and report the incident to your supervisor even if it seems minor.

TABLE 8-2

Physical Hazards (Fire)	Health Hazards
Combustible liquids	Carcinogens (cancer-causing chemicals)
Explosives	Irritants
Flammables	Corrosives
Oxidizers	Sensitizers
Reactive	Any chemical likely to adversely affect human health in any way

Placards

Placards are used to identify hazardous materials (see Figure 8-5). Placards provide information in a number of ways.

1. Color of background.
2. Symbol at top.
3. United Nations class number at the bottom.
4. Hazard class or the identification number in the center.

Placards which identify various types of hazardous materials range in various colors.

1. Orange indicates explosive.
2. Red indicates flammable.
3. Green indicates nonflammable.
4. Yellow indicates oxidizing materials.
5. White with vertical red stripes indicates flammable solid.
6. Yellow with white indicates radioactive material.
7. White over black indicates corrosive materials.

Regarding symbols which are also located on the placards, symbols suggest certain types of hazardous materials as well.

1. Bursting-ball symbol indicates explosives.
2. Flame symbol indicates a flammable.
3. Slashed W indicates dangerous when wet.
4. Skull-and-crossbones symbol indicates poisonous materials.
5. Circle with a flame indicates oxidizing materials.
6. Cylinder indicates nonflammable gas.
7. Propeller indicates radioactive.
8. Test-tube symbol indicates corrosive.
9. Word *Empty* indicates the product has been removed but the harmful residue may still be present.

Procedures

The following guidelines are recommended for protection officers who encounter a hazardous material incident:

1. Report the incident as soon as possible. Give an exact location and approach route and request assistance.
2. Stay upwind of the incident since some materials have odors and fumes that can cause injury or death.
3. Clear the area of nonessential personnel.
4. Avoid contact with liquids or fumes of any kind. Make sure the necessary protective equipment is available (gloves, masks).
5. Rescue the injured, but only if it can be done in a safe manner.
6. Identify the materials involved and determine the condition that they are in.
7. If necessary, initiate an evacuation. This should be done under management direction.

EXPLOSIVES A

POISON GAS

POISON A

FLAMMABLE

OXYGEN

FUEL OIL

FLAMMABLE SOLID

RADIOACTIVE

CORROSIVE

FIGURE 8–5   DOT placards.

8. Establish a command post so rescue efforts can be properly directed. If a hazardous condition occurs indoors, isolate and seal off the area by closing doors and shutting down the ventilation system of the area.

DISASTERS

It is impossible to predict when a disaster will occur. However, planning should be flexible enough to permit company management and the protection force to follow their own judgment in the event that a disaster occurs. The main objectives of disaster planning are to 1) save lives, 2) administer first aid, 3) minimize the loss of property and 4) return to normal operations.

It is the responsibility of management to take all possible and necessary steps to protect the interests of employees, customers, members of the public, and the property under its control. It is the responsibility of the protection officer to ensure these policies are carried out. If the company does not have a disaster plan, then one should be initiated.

There are generally two types of disasters: natural and man-made. *Natural disasters* are caused by unusual weather conditions and fires. The following is a description of some of the most common types of natural disasters.

Natural Disasters

Tornadoes. These are small, short-lived, but very violent storms with swirling winds which reach speeds up to 200–400 miles per hour. This type of storm appears as a rotating funnel-shaped cloud that extends toward the ground from the base of a thundercloud. A tornado can vary in color from gray to black and make sounds like a roaring locomotive. They are very destructive over a small area.

The public is warned by the U.S. National Weather Service when severe weather is imminent. A *severe thunderstorm watch* can signify the possibility of tornadoes, thunderstorms, frequent lightning, hail, and winds of greater than 75 mph. A *tornado watch* means a tornado may develop, and a *tornado warning* means a tornado has been sighted in the area or detected by radar. In general, a weather *watch* means possible bad weather; a *warning* means bad weather is imminent or in force. In the event of a tornado, do the following: stay inside, move to the center of the building, away from windows and doors.

Hurricanes. These storms form over the ocean with winds of 74 mph and above and are the largest storms on earth. Their force weakens when they hit coastal areas and move inland; however, flooding caused by hurricanes can result in extensive damage. Because of the storm's potential for causing damage, the National Weather Service issues hurricane advisories. If the hurricane moves toward the mainland, a *hurricane watch* is declared. If it is determined that a section of coast will be endangered, a *hurricane warning* is issued.

Forest Fires. If not quickly extinguished, small fires in wooded areas can become one of the most destructive forces caused by nature or by people. Wildlife, livestock, people, and fish can be killed; property and timber can be destroyed. Also, the ecological balance of nature can be disturbed.

Earthquakes. Usually caused by plates shifting; last a few seconds to five minutes; may be followed by aftershocks. Most injuries occur by falling debris. Procedures to follow are about the same as for a tornado. Earthquakes severity is ranked according to a scale of intensity. The *Richter Scale*, developed in 1935, is one measure of earthquake magnitude or severity. The scale is from one to nine or higher. The highest recorded earthquake in the United States was the 1906 San Francisco earthquake which indicated a magnitude of 8.25 on the Richter Scale.

Floods.   Property damage and loss of life are the results of floods. Generally, floods develop slowly; thereby allowing for adequate warning. Floods can be the result of rain and melting snow causing dams to inundate and streams and rivers to swell. *Flash floods* result when there are heavy rains in mountainous or desert areas. A *flash flood watch* alerts citizens to the potential of a flood emergency that will require immediate action; a *flash flood warning* indicates that flash flooding is occurring or is imminent and those threatened should take prompt precautions.

Winter Storms.   These vary in size, duration, and intensity and may affect many states or only part of one state. Winter storms may consist of freezing rain, sleet, or heavy snow and blizzard conditions. Storm warnings include *heavy snow* which indicates snowfall of 4 inches or more within a 12-hour period or 6 inches or more within a 24-hour period, *sleet, freezing rain, ice storm, blizzard*, and *severe blizzard warning*. Of these, a blizzard is the most dangerous, combining air, heavy snow, and strong winds of 35 mph or more and temperatures of 20 degrees Fahrenheit or lower. A severe blizzard warning has winds of at least 45 mph and temperatures of 10 degrees Fahrenheit or lower as well as heavy snowfall. The protection officer should know the road conditions, whether any roads are impassable, and which roads are being cleared.

Man-Made Disasters

Fires.   As previously discussed, fires need three components to continue burning: oxygen, fuel, and sufficient heat to ignite the fuel. If any of these three are removed, the fire will die. Fires cause direct damage by consuming valuable materials; they produce by-products that can be extremely hazardous. The primary concerns about these by-products are 1) the smoke, which may cause suffocation and temporary blindness; 2) poisonous gases, released when certain materials are burned; 3) intense heat; and 4) increased air and gas pressure which can cause an explosion if trapped in a confined area.

Firefighting methods vary making it difficult to establish standard procedures. Industrial plants, institutions, and facilities generally develop their own plans and procedures to fight various fire hazards. To effectively fight large fires, industrial explosions, and forest fires you must have good communications, accessibility to the fire scene, prearrangements for the use of equipment and human resources, and a centralized command. Foremost is the ability to respond quickly and to confine the fire before it becomes a disaster. Planning and training play a key role in fire management.

Chemical Accidents.   Chemicals by the thousands are used daily at facilities such as businesses, schools and industrial plants. If spilled, a chemical can cause an emergency affecting employees at a job site as well as others in the surrounding neighborhood. Negative effects can include contamination of the community, explosions, and fire.

Transportation Accidents.   Because almost every facility or establishment is exposed to the possibility of air, automobile, rail, or shipping accidents within or near its boundaries, it is important that security and safety personnel be prepared to handle problems resulting from a major transportation accident. Knowledge of special rescue and evacuation procedures are critical when handling major transportation accidents resulting in chemical spills, fires, and contamination of the air. Regardless of the type of transportation accident, saving lives is always the first and most important consideration.

Civil Disturbances.   Public demonstrations may develop slowly and be nonviolent, or rapidly and with little warning, they may turn violent. Labor strikes often develop slowly and protection personnel are called in as needed. If violence occurs, protection personnel should be involved from the start and can act as the primary source of information about the nature and extent of the disturbance.

Sabotage.   In general, sabotage is committed by disgruntled employees for revenge or to further labor causes, or by persons politically motivated or mentally ill. Methods of sabotage include:

1. *Chemical.* Involves the addition of destructive or polluting chemicals.
2. *Electrical or electronic.* Involves the interruption of or interference with electrical processes, including the jamming of communications.
3. *Explosive.* Damage or destruction of machinery using explosives.
4. *Incendiary.* Ignition of fires by chemical, electrical, electronic, or mechanical means.
5. *Mechanical.* Destruction, removal or omission of parts, use of improper or inferior parts, or the failure to lubricate or maintain equipment.
6. *Psychological.* Instigation of strikes, boycotts, unrest, personal animosities, and other actions that result in work slowdowns.

Sabotage threats can be reduced by making targets less accessible and vulnerable and establishing policies and procedures for handling potential or actual sabotage.

Nuclear Attack.   Probable targets are military installations, industrial centers, and large cities. Though unlikely to occur, the protection officer must be aware of the threat and be able to react to the effects of the blast and heat (thermal radiation).

Radiological Accidents.   The possibility of radiological accidents has increased since the advent of widespread and rapidly increasing use and transportation of radioactive materials. When accidents of this type occur, some response is immediately required by employees, local police, and fire personnel. State and federal action generally follows.

These are but a few examples of the types of disasters that can occur in the company or organization. Some companies should include preparations for possible enemy or terrorist attacks by bombing in their disaster planning.

In the event that there is a disaster, it should not be assumed that local police and fire services will be able to assist with every emergency. In the case of a widespread disaster, the police and fire organizations might not be able to respond because of impassible roads. Or, if many facilities need assistance at the same time, sufficient personnel may not be available to all organizations needing help. Therefore, provisions should be made in disaster planning for additional employees to assist those regularly assigned to security and fire-protection duties.

Because fire is the single greatest destroyer of property and generally accompanies almost any type of disaster condition, an emergency fire protection group should be organized and trained to help limit damage in any emergency.

As a protection officer, you may be asked to become the person trained in the area of fire protection. Local fire organizations usually have personnel available to conduct training classes on how to provide such services.

Guidelines for Disaster Response

As discussed, when disaster occurs there should be a disaster plan prepared by management. All protection officers should be trained what to do in case of disaster. Following are a number of guidelines that should be followed in the event of a disaster:

1. Set up emergency headquarters as quickly as possible. A command post of some type should be set up by security management to enable protection officers to respond to the disaster quickly.

2. Evacuate areas where dangers may exist (inside buildings, fallen power lines).

3. Notify the fire department, police department, and company management of the type of disaster that has occurred. Do not assume that notifications have already been sent; notifications are critical in the case of a disaster. If standard communication lines are down, be prepared to use alternate means of notification (for example, wireless transmission).

4. Request medical assistance. Make sure that those in need of medical assistance receive attention immediately. Give the location and the type of medical equipment needed.

5. Look out for pilferage, looting, and other types of uncontrolled violence that frequently occurs during periods of disaster.

6. If the situation permits, inspect building interiors and other areas to make sure that nobody has been trapped inside (restrooms, elevators). Do not enter a building or other structure if it is unsafe to do so. Keep all unauthorized people away from the disaster areas. Only authorized persons such as rescue persons should be allowed inside the disaster area.

7. Record all incidents that you have observed (looting, other types of pilferage). Make a written note of all types of problems observed during the disaster.

8. Know where all emergency first-aid equipment is located. This includes batteries, portable radios, blankets, and other types of equipment needed during a disaster.

9. Remain calm and do not panic. When a disaster occurs, people will be looking to protection officers as a source of guidance. Unless properly relieved, remain at your post. Maintain constant communication with superiors or other protection officers to determine what your duties may be; you may be sent to a different location to assist others.

Although it is the responsibility of security management to provide emergency and safety supplies, the protection officer should make recommendations to management concerning missing items. The following list of emergency and safety devices are recommended for large facilities which have a large employee population and/or a large population of visitors that may be in the facility at the time that a disaster would occur. Such places would be possibly large industrial plants, schools, manufacturing plants, or entertainment complexes. Emergency and safety devices should include, but not be limited to, the following:

1. Flares
2. First aid kits

3. Oxygen
4. Spot lights
5. Flash lights
6. Signal lights and flares
7. Fire blankets
8. Fire extinguishers
9. Matches
10. Fuels (wood and petroleum)
11. Life preserver equipment, if necessary
12. Emergency food packs
13. Emergency drinking water
14. Emergency washing water
15. Generators
16. Emergency medical packs

The previous list is not exhaustive but should serve as a guide as to the equipment needed in the case of disaster. Emergency equipment should be purchased only after a careful risk assessment has been completed for the facility.

FIRST AID AND CPR

In situations involving illness or injury to others, the protection officer may be the first to respond to the scene (see Figure 8–6). Although professional rescuers have many different occupations, they share two important characteristics. First, they have the duty when they are on the job to respond to an emergency. Second, they have been professionally trained to use certain techniques that are not generally taught in CPR courses offered to the general public. A protection officer can be considered a professional rescuer.

FIGURE 8–6

It is important that a protection officer be trained in the principles of first aid and CPR. However, it must be recognized that the first aid and CPR training covered in this section of the chapter will not replace, nor is it meant to replace, a certified course in first aid and CPR. It is highly recommended that protection officers attend a certified course in first aid and CPR from a local chapter of the Red Cross. It is essential to learn the basic skills of first aid and the more advanced skills of CPR. The material discussed in this section is taken from the *Standard Red Cross First Aid and CPR Training Guide* (figures depicted courtesy of the American Red Cross).

The Law

Legally, a victim must give consent to an offer of help before a person trained in first aid begins to help him or her. The law assumes that an unconscious person would give consent. If a person is conscious, the protection officer must ask for permission before rendering first aid. Protection officers should make a reasonable attempt to get consent from any parent or guardian of a victim who may appear to be a minor or who appears to be mentally or emotionally disturbed. However, if the parent or guardian is not available, the protection officer may give first aid without consent. Consent is implied for a person who is unconscious, badly injured, or so ill that he or she cannot respond. Many states have good-Samaritan laws which give legal protection to rescuers such as protection officers who act in good faith and are not guilty of negligence or willful misconduct. It is recommended that you check your state and local statutes to better understand good-Samaritan laws.

Infectious Disease Precautions

When rendering first aid, it is important to adopt practices that prevent the spread of infectious diseases. Of concern are blood-borne diseases such as hepatitis and HIV (the AIDS virus) and air-borne diseases such as influenza and tuberculosis. The AIDS virus can also be transmitted by drug users sharing needles, so all needles should be handled with care.

To minimize the risk of infection when attempting to control bleeding, use some sort of barrier such as dressings, latex gloves, or a piece of plastic wrap between you and the victim's blood. Also, wearing latex gloves prevents direct skin contact with other body fluids such as vomit, feces, or urine. After giving first aid, wash your hands as soon as possible.

Mouth-to-mouth contact in CPR and rescue breathing is another source of possible infectious disease transmission. However, to date there has been no evidence to show that the Hepatitis B virus or HIV (AIDS virus) can be transmitted either through contact with human saliva or by giving rescue breathing. Also, according to the Center for Disease Control there has never been a documented case of any infectious disease transmitted through the use of CPR manikins.

First Aid

When responding to an emergency, it is important to decide what course of action to take. There are recommended steps that must be followed in responding to a situation involving an injury or accident.

Survey the scene. The protection officer must decide if the situation is safe by surveying the scene. You cannot help a victim if you become a victim. Avoid running into burning buildings or exposing yourself to electric shock. Surveying the scene also means determining what happened or what was the cause of the accident or injury. If the victim is conscious, ask specific questions to determine what happened. Try to determine the extent of the victim's illness or injury. If the victim is unconscious and you are unable to determine what caused the illness or injury, look for clues. The scene itself may give you some indication as to how the person was injured (broken glass, electric wires, fumes). Quickly look for medical tags on the victim's neck or wrist because these tags may indicate if the person is diabetic or epileptic. Also be concerned about accidents which may have caused head, neck, or back injuries. If the injury is not treated properly, it could lead to paralysis, or even death. Therefore, if the victim complains of pain in the neck, head, or back or is found unconscious after an accident, consider the possibility of a spinal injury. Do not move the victim until responsible medical assistance has arrived.

Surveying the scene also means determining how many people are injured. Some victims may be difficult to find. Make certain no one is overlooked. If there are bystanders, use them to find out what happened. If anyone knows the victim, ask if the victim has any medical problems. If the injury or illness occurs on company property, co-workers and supervisors may know something about the victim's background.

Survey the victim. After surveying the scene, the next step is to survey the victim. The purpose of surveying the victim is to check for life-threatening conditions and to administer first aid when necessary. Some illnesses or injuries may require assistance in caring for the victim. In this case, you may need to shout several times to get someone's attention. While you are seeking help, do an ABC check of the victim.

A. Check the airway. Does the victim have an open airway? Is the victim breathing? The most important act for a successful resuscitation is to immediately open the unconscious victim's airway using the head-tilt, chin-lift method (Figure 8–7). This position lifts the tongue away from the back of the throat and opens the airway. Detailed steps are given in the Red Cross training course.

B. Check for breathing. Look for the chest to rise and fall, listen for breathing, and feel for air coming out through the victim's nose. If the victim is breathing, you will see chest movement and hear and feel escaping air.

C. Check for circulation. Is the person's heart beating? If a person is breathing, his or her heart is beating and circulating blood. If a person is not breathing, you must find out if the heart is beating. This is accomplished by checking for his or her pulse. Feel for a pulse on the side of the neck, also on the wrist (Figure 8–8). Make sure to complete the primary survey before giving any urgent first aid. Life-threatening conditions must be cared for before less serious conditions

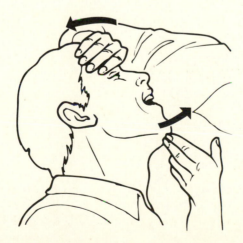

FIGURE 8–7

are treated. It is more important, for example, to give rescue breathing to someone who is not breathing than to splint a broken arm or bandage a small cut. If no one has responded to your shouts for help by the end of one minute of giving first aid, get to a phone as quickly as you can and phone for emergency medical assistance. Then return quickly to the victim and continue first aid.

Phone for emergency medical services.　The third principle of emergency action is phoning for emergency medical service. The emergency-medical-service system is a community-coordinated means of responding to an accident or medical emergency. If you work in an industrial setting, contact the hospital or health-care facility at your location. Otherwise, contact the community emergency medical system. Once you have completed a primary survey of the victim, you can activate the professional system that provides advanced care and skilled transfer of the victim to a medical facility. Contact emergency medical services (EMS) when any of the following conditions exist:

1. Unconsciousness or altered level of consciousness
2. Breathing problems (difficulty in breathing or no breathing)
3. Persistent chest pain or abdominal pain or pressure
4. No pulse
5. Severe bleeding
6. Vomiting blood or passing blood
7. Poisoning
8. Seizures, severe headaches, or slurred speech
9. Injuries to head, neck, or back
10. Possible broken bones

There are also special situations that warrant calling EMS personnel for assistance. These include

1. Fire or explosion
2. The presence of poisonous gas
3. Downed electrical wires
4. Swiftly moving water
5. Motor vehicle accidents
6. Victims who cannot be moved easily

FIGURE 8-8

Of course, when in doubt, always contact emergency medical services as well as the appropriate supervisors when there is an illness or injury.

First Aid Response to Special Emergencies

There are a number of situations in which a protection officer may need to use first aid skills. The following areas will be discussed: rescue breathing; choking; bleeding; shock; poisoning; burns; fractures; dislocations; sprains, and strains. Heart attack and heart arrest will be discussed in the section on CPR.

Rescue Breathing. Breathing emergencies may be caused by a number of factors such as air obstructions, poisonous substances, injuries to the chest or lungs, near-drowning, electrocution, drugs, burns, certain diseases and illnesses, shock, and reaction to certain insect bites and stings.

When you encounter a person who may have breathing difficulties, do an ABC (open airway, check breathing and circulation) check.

1. Check for unresponsiveness.
2. If no response, shout for help.
3. Position the victim on his or her back.
4. Open the airway.
5. Listen, look, and feel for breathing.
6. If the person is not breathing, give two full breaths into the mouth.
7. Check the carotid (side of the neck) pulse; and look for severe bleeding.
8. Have someone phone for emergency medical services (EMS).

Rescue breathing is a way of breathing air into someone's lungs when natural breathing has stopped, or a person is unable to breathe properly on his or her own. This is also known as the *artificial respiration* method. It is important to remember that in the absence of a pulse, a person cannot be breathing. When the respiratory system shuts down, the heart stops beating and the person then stops breathing.

Rescue breathing is administered when the victim is not breathing but still has a pulse. Position by placing victim on his or her back and open the airway with the head-tilt/chin-lift technique and give two full breaths; then give one breath every five seconds. Each breath should last one to one and one-half seconds. After one minute, recheck the carotid pulse and then continue giving one breath every five seconds. Recheck pulse every minute. Have someone phone for EMS help.

Choking. Choking is also known as *airway obstruction*. It occurs when the airway becomes blocked by some type of object, fluids, or the back of the tongue. The person who is choking may quickly stop breathing and lose consciousness. Most choking is caused by 1) trying to swallow large pieces of food that are improperly chewed, 2) drinking alcohol before or during eating, or 3) wearing dentures. Dentures make it difficult to sense the size of food when chewing and swallowing. Other causes of choking include talking excitedly or laughing while eating; eating too fast; and walking, playing, or running with objects in the mouth.

Some of the signals to look for in choking emergencies would be when a person is coughing loudly, and is unable to speak, or he or she may also wheeze between breaths, or the person may also clutch at his or her throat with one or both hands. This is universally recognized as a distress signal for choking (Figure 8–9). If the person is coughing or wheezing, do not interfere with his or

FIGURE 8–9

her attempts to cough up the object. Stay with the person and encourage his or her efforts to continue coughing. If coughing persists, call EMS for help.

If the victim is conscious, ask them "Are you choking?" If he or she is choking then perform the *Heimlich Maneuver* until the obstruction is cleared or the victim becomes unconscious. This maneuver is accomplished by 1) getting behind the victim and wrapping your arms around the waist (Figure 8–10); 2) making a fist with one hand and placing it with the thumb knuckle pressing inward, just below the point of the *v* of the rib cage; 3) grasping the wrist with the other hand and giving one or more upward thrusts or hugs (Figure 8–11), and 4) starting mouth-to-mouth if breathing has stopped. In case of an unconscious victim, find out if the unconscious victim has an airway obstruction. Begin with the primary survey to check the ABCs as you did for rescue breathing.

1. Check for unresponsiveness.
2. If there is no response, shout for help.
3. Position the victim on his or her back.
4. Open the airway.
5 Check for breathlessness.
6. If no breath, give two full breaths into the mouth.
7. If you are unable to breathe air into the victim, retilt the victim's head and give two full breaths again.
8. Have someone phone for emergency medical services.
9. Perform 6–10 Heimlich maneuvers.
10. Do a finger sweep.
11. Give two full breaths.

Repeat the last three steps until the obstruction is clear or help arrives.

Finger Sweep.   If a person who is choking becomes unconscious during the Heimlich maneuver, shout for help and slowly lower the person to the floor. Open the victim's mouth, holding the tongue and lower jaw between the thumb and fingers, and lift. This will move the tongue from the back of the throat. The index finger can then "sweep" the area for any foreign object. Caution: A partial obstruction may become a total obstruction if you push the object farther down the throat. The other danger is that the victim may bite your fingers.

FIGURE 8-10

Bleeding. Bleeding can be external or internal. Arterial bleeding is the most serious because it is life threatening. In cases of severe or uncontrolled bleeding, call the ambulance as soon as possible. Shock and loss of consciousness may occur from the rapid loss of as little as a quart of blood. Because it is possible to bleed to death in a very short period of time, immediate first aid must be taken. Signs of life-threatening external bleeding are: 1) blood spurting from the wound, and 2) bleeding that won't stop after all measures have been taken to control it.

What To Do To Control External Bleeding

1. Apply direct pressure to the wound.
2. Have someone call a hospital ambulance.
3. Elevate the wound.
4. Treat the victim for shock.
5. Monitor the ABCs.

FIGURE 8-11

Direct pressure over the site of the wound is the best way to control bleeding. Use the palm of your hand if a clean cloth, diaper, or pad is unavailable. Apply direct pressure to the area—press hard. Do not look under the pad to see if bleeding has stopped. Most bleeding will stop within a few minutes. If pad becomes blood-soaked, do not remove; instead, add additional layers of cloth and increase pressure. Unless a fracture is suspected, elevate the wound above the level of the victim's heart and continue to apply pressure. Continue to monitor the ABCs. A tourniquet is no longer used because it does more harm than good.

Internal bleeding is harder to recognize than external bleeding. Symptoms of internal bleeding include:

- Bruising or discolored skin in the injured area
- Soft tissues, such as those in the abdomen, that are tender, swollen, or hard.
- Anxiety or restlessness.
- Fast, weak pulse.
- Fast breathing.
- Skin that feels cool or moist, or looks pale or bluish.
- Nausea and vomiting.
- Strong thirst.
- Decrease in the level of consciousness.

If you suspect that the victim is suffering from a severe internal injury and experiencing internal bleeding, call EMS immediately.

WHAT TO DO TO CONTROL INTERNAL BLEEDING (WHILE WAITING FOR EMS):

1. Do no further harm.
2. Check the ABCs.
3. Help the victim rest in the most comfortable position.
4. Maintain normal body temperature.
5. Reassure the victim.
6. Care for other injuries or conditions.

Shock. The condition known as *shock* results from a depressed state of many vital body functions which could be life-threatening even though the victim's injuries would not otherwise be fatal. When vital organs such as the brain, heart, and lungs do not receive oxygen-rich blood, they do not function properly; this results in shock. Shock is characterized by restlessness or irritability; pale, cold skin; a rapid but weak pulse; nausea or severe thirst and drowsiness; or loss of consciousness. In later stages, the victim becomes unresponsive and pupils dilate. Remember, shock can be present even without the visible evidence of any injury.

If the symptoms of shock are present, do the following:

1. Check the ABCs.
2. Give the necessary first aid immediately to eliminate the causes of shock such as excessive bleeding or stoppage of breathing.
3. Have victim lie flat (Figure 8–12) and cover him with a light blanket. Preferably the victim should not be lying down on the bare ground; keep warm enough to avoid or overcome chilling, but do not overheat. React to weather conditions.

FIGURE 8-12

4. Make sure the victim has an open airway. If victim is vomiting, place on one side and let fluids drain from mouth.
5. If the person has trouble breathing, place him or her into a semireclining position to make breathing easier. Use boxes, pillows or blankets to raise the head and back (Figure 8–13).
6. If there are no head injuries, elevate the legs slightly as shown in Figure 8–14 (8 to 12 inches above the head).
7. If there are head injuries, keep head higher than rest of body.
8. Call an ambulance immediately.

Remember to give the care that may help save the victim's life, you do not have to identify what is wrong. If you recognize the signals of shock, give first aid immediately.

Poisoning. *Poison* can be a solid, liquid, or gas that causes injury or death when introduced into the body. The four main ways a person can be poisoned are by swallowing, inhaling, absorbing through the skin, and by injection (Figure 8–15).

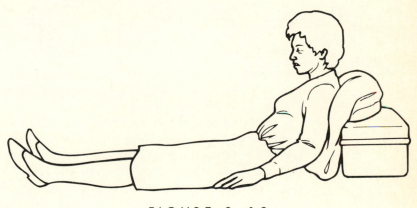

FIGURE 8-13

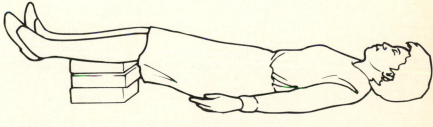

FIGURE 8-14

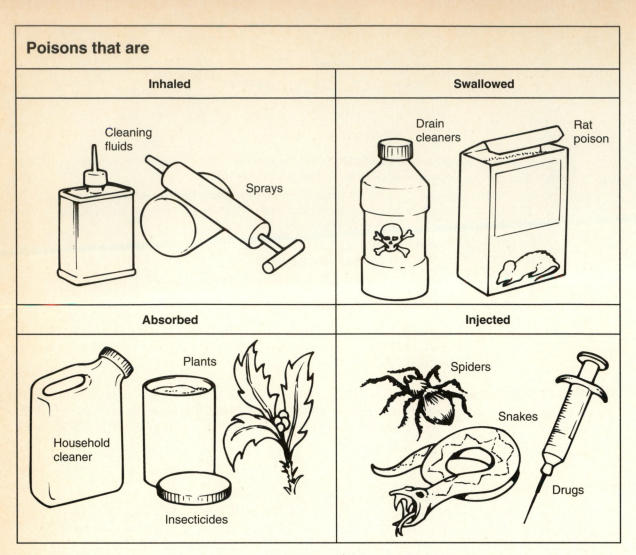

Poisons that are

Inhaled	Swallowed
Cleaning fluids / Sprays	Drain cleaners / Rat poison

Absorbed	Injected
Household cleaner / Plants / Insecticides	Spiders / Snakes / Drugs

F I G U R E   8 – 1 5

EMS personnel face some of the same problems of the lay public in that some poisons are quick-acting while others act slowly and cannot easily be identified. Therefore, the most important thing is to recognize that a poisoning may have occurred. First, survey the scene and then do a survey of the victim. Try to establish whether poison was absorbed, swallowed, or inhaled.

Steps to follow if victim swallowed poison:

1. Place victim on side if vomiting.
2. Phone EMS and the poison control center; have containers in hand if possible.
3. Follow directions from EMS and the poison control center.
4. Monitor ABCs.
5. Save containers and any vomitus to give EMS.

Steps to follow if victim inhaled poison:

1. Shout for help.
2. Remove victim from source of the poison.

3. Get victim fresh air.
4. Do a primary survey of victim.
5. Place victim on side if vomiting.
6. Phone EMS and poison control center.
7. Follow their directions.
8. Monitor ABCs.

Steps to follow if victim absorbed poison:

1. Remove victim from the source of the poison.
2. Wash poison from skin.
3. Remove clothing and other articles with poison on them.
4. Phone EMS and the poison control center.
5. Follow their directions.
6. Monitor ABCs.

Burns.   A protection officer will most likely be concerned with heat burns, electrical burns, and chemical burns. Burns are classified according to their source (heat or chemicals) and their depth. The three depth classifications include *first-degree* (superficial) *second-degree* (partial thickness), and *third-degree* (full thickness) burns.

When you encounter a situation where persons have been critically burned, immediately call EMS for assistance. Critical burns are potentially life threatening, disfiguring, or disabling. Call EMS in the following situations:

* Burns whose victims are having trouble breathing.
* Burns that cover more than one body part.
* Burns on the head, neck, hands, feet, or genitals.
* Any second-degree or third-degree burn to a child or elderly person.
* Burns resulting from chemicals, explosions, or electricity.

Fractures, Dislocations, Sprains, and Strains.   Fractures, Dislocations, sprains, and strains are musculoskeletal-system injuries. This system consists of bones, muscles, ligaments, and tendons.

Fractures.   Fractures are classified as open or closed and are breaks or cracks in bones. In a closed fracture the skin is unbroken; in an open fracture you encounter an open wound. The open fracture is the more serious because of the risk of infection and severe bleeding. Severe shock can accompany the fracture of a large bone. Fractures can also be accompanied by internal injuries. For example, a person can have fractured ribs and internal injuries to the lungs, kidneys, or liver.

Dislocations.   These are injuries in which a bone is separated or displaced from its normal position at a joint. Ligament damage may be involved.

Sprains.   This injury is a partial or complete tearing of ligaments and other tissues at a joint. The more ligaments that are torn, the more severe the injury. Pack in ice to keep swelling down.

Strains.   A strain is caused by a stretching or tearing of muscle or tendon fibers. Pack in ice to keep swelling down.

When there is a question as to whether an injury is a fracture, dislocation, sprain, or strain, treat the injury as you would for a fracture. If EMS is on its way control bleeding but do not move the victim. Treat for shock and monitor ABCs. As a protection officer your main concerns are to

- Survey the scene.
- Survey of the victim.
- Phone EMS if necessary.
- Do a secondary survey.
- Establish if injury was caused by force.

If you suspect a head or neck injury, do not move the victim or stop the flow of blood or clear fluids coming from the nose or ears.

Cardiopulmonary Resuscitation (CPR)

Recognizing the signals of a heart attack or cardiac arrest can prevent death. Heart attacks are characterized by chest discomfort or pain. The pain may be described as uncomfortable pressure, squeezing, a fullness or tightness, aching, crushing, constricting, oppressive, or heavy. Other symptoms include sweating, nausea, and shortness of breath. Heart attacks usually occur in the morning hours but can happen at any time. Any heart attack can lead to cardiac arrest (stopping of the heart) so prompt action can save a life.

As a protection officer, you should be able to recognize the signals of a heart attack and take appropriate action.

1. Have the victim stop what he or she is doing and have them sit or lie down in a comfortable position. Do not let the victim move around; loosen any restrictive clothing the victim might be wearing.
2. Ask someone to phone EMS for help (if you are alone, make the call yourself).
3. After calling EMS, survey the victim to get information about his or her condition. Bystanders may be able to give you information.

Information about the victim should include the victim's name, age, previous medical problems, and a description of the duration and type of pain (for example, dull, heavy, or sharp). Be prepared to give CPR if the heart stops (cardiac arrest occurs).

CPR is a combination of artificial respiration and artificial circulation. Artificial circulation (usually called *external heart compression* or *external heart massage*) must only be applied by trained rescuers (Arnold, 1980). CPR must be started as soon as possible after the heart stops.

Step-by-step technique. Check the ABCs as in rescue breathing.

1. Determine if the victim is unconscious by gently shaking him and shouting "Are you OK?" If you get no response, shout for help.

    Place victim flat on his back on a hard flat surface. Then open the airway by tilting the head backward and lifting the neck up from behind until the chin points straight up (Figure 8–16).

    Place your ear at the victim's mouth and nose; watch the chest for breathing movements. Look, listen, and feel for breathing activity. Check for breaths against your cheek. If signs are absent, proceed immediately to the next stage of CPR.

2. Pinch nose closed. Place your mouth tightly around the victim's mouth (Figure 8–17).

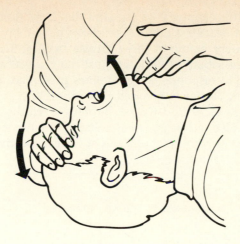

FIGURE 8–16

Give two quick full breaths. Then check the carotid pulse and breathing for five to ten seconds. Check for pulse near the Adam's apple.

If there is a pulse, but no breathing, give one breath every five seconds; if there is no pulse or breathing, restate the need for emergency help and begin chest compressions.

Place the heel of one hand on the notch where the ribs meet the breastbone. Clasp your other hand on the first hand, keeping fingers off the chest.

3. When doing chest compressions, push straight down without bending your elbows. Shoulders should be over the victim's breastbone. Push down about one and one-half to two inches.

Watch the victim's chest to see that it falls; rhythmically repeat the blowing cycle, once very five seconds or twelve times per minute. Give fifteen compressions at a rate of eighty per minute and give two full breaths. Check pulse and breathing after the first minute and every few minutes thereafter. If performing two-person CPR, give one breath for every five compressions.

Remember, do not begin CPR if the victim has a very slow or very weak pulse. It is not necessary to bare the victim's chest. It is dangerous to do chest compressions if the victim's heart is beating so it is important to check the carotid pulse for five to ten seconds before you start CPR. If the victim's clothing interferes with finding the proper location for chest compressions, bare only a small area, enough for hand placement. Continue administering CPR until the heart starts beating again, a second rescuer trained in CPR takes over for you, EMS personnel arrive and take over, or you are too exhausted to continue.

FIGURE 8–17

Conclusion The purpose of this chapter is to acquaint the protection officer with health and safety concerns. Regardless of the type of business, there are a number of threats to health and safety. Some of these threats may be accidental; others may be intentional acts of sabotage. Officers must not only recognize these threats, but also be prepared to respond. The guidelines presented must be understood and practiced. It is highly recommended that officers become certified in First Aid and CPR through local Red Cross chapters. Periodic in-service training and recertification is also necessary. Keep a list of emergency phone numbers available (Figure 8–18) in your notebook and at your respective post or office.

EMERGENCY PHONE NUMBERS

FIRE DEPARTMENT: _____ PHONE: _____
                          Address

POLICE DEPARTMENT: _____ PHONE: _____
                          Address—Local

POLICE DEPARTMENT: _____ PHONE: _____
                          Address—State

REMEMBER—If you can't reach a doctor, dial 911 and tell the operator that you have an emergency. Be prepared to give the correct address.

DOCTOR: _____ PHONE: _____
                 Address

AMBULANCE: _____ PHONE: _____
                   Address

PARAMEDICS: _____ PHONE: _____
                    Address

HOSPITAL: _____ PHONE: _____
                  Address

CARDIAC UNIT: _____ PHONE: _____
                      Address

POISON CONTROL CENTER: _____ PHONE: _____
                               Address

PHARMACY: _____ PHONE: _____
                  Address

TAXICAB: _____ PHONE: _____
                 Address

FIGURE 8–18

1. Explain the purpose of OSHA.
2. What is the fire triangle?
3. What are the four classifications of fire?
4. What type of fire extinguisher is recommended for the following fires?
   - Wood fires
   - Gasoline fires
   - Electrical fires
5. List a protection officer's responsibilities in the event of receiving a bomb threat by phone.
6. What is the purpose of hazardous materials placards?
7. What are good-Samaritan laws? Does your state have such laws?
8. In responding to an injury or accident requiring first aid, explain the process of surveying the scene, surveying the victim, and phoning for emergency medical assistance.
9. How is the AIDS virus transmitted?
10. In the following types of medical emergencies, explain the recommended Red Cross response.
    Choking
    Arterial (external) bleeding
    Heart attack
    Shock
    Suspected poisoning
    Internal bleeding
    Fractures, dislocations, sprains, and strains
    Suspected head or neck injury
11. What emergency phone numbers should be kept by the protection officer and/or protection dispatcher?

References

AMERICAN RED CROSS, *American Red Cross Standard First Aid Workbook*, The American National Red Cross, 1991.

AMERICAN RED CROSS, *Standard First Aid & Personal Safety* (2nd ed.). New York: Doubleday & Company, 1980.

AMERICAN RED CROSS, *The Emergency Survival Handbook*. Los Angeles Chapter, American Red Cross, Los Angeles, CA, 1985.

ARNOLD, PETER, WITH EDWARD L. PENDAGAST JR., M.D., *Emergency Handbook*, A First-Aid Manual for Homes and Travel. New York, NY: Doubleday & Company, 1980.

CITE, ARTHUR, AND JIM L. LINVILLE, EDS., *Fire Protection Handbook* (16th ed.) Quincy, MA: National Fire Protection Association, 1986.

COAKLEY, DEIRDRE, *The Day the MGM Grand Hotel Burned*. Secaucus, NJ: Carrol Publishing Group, 1989.

FISCHER, ROBERT J., AND GION GREEN, *Introduction to Security* (5th ed.). Stoneham, MA; Bunsaworth-Heinemann, 1992.

Substance Abuse

Learning
Objectives

After studying this chapter, you should be able to explain the following:

Alcohol abuse
Barbiturates
Cannabis
CNS stimulants
Depressants
Drug dependency
Drug slang
Drug screening programs
Hallucinogens

Inhalants
Marijuana
Narcotic substances
Narcotics
Drug Addiction
Recognizing the drug abuser
Stimulants
Uniformed Controlled Substances Act

INTRODUCTION

One of the most serious problems facing our society is drug abuse. Although some believe that drug abuse should be treated as a health problem, the abuse of drugs is often associated with crime. Drugs include alcohol, marijuana, cocaine, and other mind-altering substances which affect a person's behavior and his or her relationship with family, friends, and employers.

Since drug abuse can be a problem in the work place, a protection officer must be prepared to recognize the warning signs of persons abusing a substance. There can be serious security and safety concerns if employees or invitees are permitted to operate in a business environment while under the influence of alcohol or other drugs. This section will provide the protection officer with some basic information regarding the law, the types of drugs abused, and response strategies.

THE UNIFORMED CONTROLLED SUBSTANCES ACT

All states have laws regulating illegal drug use. There are laws against illegal possession, manufacture, sale, and distribution of drugs. It is important to review local laws in order to gain an understanding of the drug laws in your jurisdiction. One major law with which the protection officer should be familiar is the Uniformed Controlled Substances Act enacted by all the states and the federal government. The intent of this act has been expressed as follows:

> This Uniform Act was drafted to achieve uniformity between the laws of the several States and those of the Federal government. It has been designed to complement the new Federal narcotic and dangerous drug legislation and provide an interlocking trellis of Federal and State law to enable government at all levels to control more effectively the drug abuse problem.[*]

Pursuant to this law, it is a criminal offense in most states to

- Manufacture or deliver a controlled (forbidden) substance.
- Possess with intent to manufacture or deliver a controlled substance.
- Create, deliver, or possess with intent to deliver a counterfeit substance.
- Offer or agree to deliver a controlled substance and then deliver or dispense a substance that is not a controlled substance,
- Knowingly keep or maintain a store, dwelling, building, vehicle, boat, or aircraft, etc. resorted to by persons illegally using controlled substances.
- Acquire or obtain possession of a controlled substance by misrepresentation, fraud, forgery, deception, or subterfuge.

There are five schedules of controlled substances in the Uniform Controlled Substances Act. The drugs are categorized by potential abuse, the degree to which they are accepted for medical use, and their relative physical danger to the abuser. These schedules are summarized as follows:

Schedule I. The substances, both narcotics and nonnarcotics, in this schedule have
(a) A high potential for abuse.

---

[*]Uniform Controlled Substances Act, 9 U.L.A. Commissioners' Prefatory Note, 1979, p.188.

(b) No currently accepted medical use.

(c) No acceptable safety standards for supervised medical use.

This schedule includes drugs such as illegal narcotics and hallucinogens such as opiates heroin, LSD, and mescaline. The maximum possible imprisonment for the manufacture or delivery of the narcotic substances is listed in this schedule. Money fines may also be imposed. The penalties for the manufacture or delivery of the nonnarcotic substances are also listed in this schedule.

Schedule II. This schedule contains both narcotic and nonnarcotic substances which

(a) Have a high potential for abuse.

(b) Have currently accepted medical uses, even though some uses may have severe restrictions.

(c) May lead to severe psychological or physical dependence.

This schedule includes drugs such as opium, coca leaves and their derivatives, opiates (synthetic narcotics such as methadone), amphetamine, methamphetamine, methylphenidate, and phenmetrazine. The maximum imprisonment possible for the illegal manufacture or delivery of the substances in this schedule is often identical to Schedule I.

Schedule III. This schedule contains substances which

(a) Have a potential for abuse less than the substances listed in Schedules I and II.

(b) Have a currently accepted medical use.

(c) May lead to moderate or low physical dependence or high psychological dependence.

This schedule includes stimulants such as amphetamines; depressants, such as short-acting barbiturates; and certain narcotic combinations such as codeine preparations. Penalties are listed in the schedule.

Schedule IV. This schedule contains substances which

(a) Have a low potential for abuse relative to the substances in Schedule III.

(b) Have a currently accepted medical use.

(c) May lead to limited physical or psychological dependence relative to the substances in Schedule III.

This schedule includes long-acting barbiturates such as phenobarbital, and minor tranquilizers such as meprobamate. The maximum imprisonment possible for the illegal manufacture or delivery of the substances is listed in this schedule.

Schedule V. This schedule contains substances which

(a) Have a low potential for abuse relative to the substances in Schedule IV.

(b) Have currently accepted medical use.

(c) May lead to limited physical or psychological dependence relative to the substances in Schedule IV.

This schedule includes narcotic drugs containing nonnarcotic active medical ingredients, and certain stimulants or depressants containing active medical ingredients. The maximum imprisonment possible for the illegal manufacture or delivery of the substances is listed in this schedule.

Drug use and sales have increased criminality in a number of ways. Because some drugs cause dependence and are illegal, they can be expensive habits to maintain. As a result, drug-dependent persons often commit crimes to support their habits. There are several definitions associated with drug abuse. *Addiction* refers to physical change, tolerance, and the presence of withdrawal symptoms; whereas *drug dependency* is a physical or psychological need for any type of drug. Researchers suggest that a majority of crimes such as burglary, robbery and other thefts are committed by drug-dependent people. For example, between 45 percent and 82 percent of the males arrested for violent crimes in selected major cities tested positive for drugs (U.S. Department of Justice, 1989:8).

If the person is employed, it is rare that the abuser will make enough money to support a habit that may exceed $200 per day. As a result, many innocent persons may be victimized by many of these persons who are dangerous and desperate who will stop at nothing to get money or property to support their habits. Employee theft may increase as well. This is especially true in businesses that have a high cash flow, or deal in large quantities and/or movements of merchandise (shipping and receiving). According to a college poll, one in four American workers has personal knowledge of co-workers using illegal drugs on the job (Kelly, 1989).

The federal government has estimated that reduced productivity due to alcohol and drug abuse costs the United States $100 billion annually. Drug use on the job can result in accidents, absenteeism, health problems, money loss, turnover and loss of coordination and skills. The problem has been so serious that among the Fortune 500 companies (largest American corporations), 50 percent are estimated to use drug-screening programs. Alcohol and marijuana are the most commonly abused drugs in the workplace (Liska, 1990).

The implications of drug abuse in the workplace are clear. Protection officers are going to face greater demands for ever-increasing involvement in drug-abuse surveillance and enforcement duties.

WARNING SIGNS OF DRUG DEPENDENT PERSONS

There are a number of indications suggesting that a person may be abusing drugs. While employees who are abusing drugs may be more of a "management problem," you may be in a position to initially detect or observe such persons. There may be a number of reasons why a person uses drugs. Many suggest certain socioeconomic and political conditions, others argue boredom and the need for excitement are reasons for drug abuse. Whatever the reason, other persons may be at risk if an employee is operating equipment or making decisions relating to safety. Drug-dependent persons frequently become unfit for employment as their mental, emotional, and physical conditions deteriorate. The following lists possible symptoms of drug-dependent persons:

- Frequent absenteeism from job.
- Sudden and dramatic changes in discipline and job performance.
- Unusual degrees of activity or inactivity.
- Sudden and irrational flareups.
- Significant change in personal appearance for the worse.
- Dilated pupils or wearing sunglasses at inappropriate times or places.
- Needle marks, razor cuts, or long sleeves constantly worn to hide such marks.

- Sudden attempts to borrow money or to steal.
- Frequent association with known drug abusers or pushers.
- An appearance of drowsiness, sleepiness, or lethargy.

CATEGORIES OF DRUGS

There are a number of categories of drugs. Some drugs are legal to possess while others are illegal or require a prescription. Drugs are derived from many sources such as plants (cocaine), animals (tetanus), laboratory synthesis (aspirin), microorganisms (penicillin), and DNA or genetic finger-printing technology (insulin). Drugs affect people in different ways. Some drugs cause violent unpredictable behavior, depression or illness. Drugs, including alcohol, can cause death if abused. Generally, there are eight categories of drugs (Table 9–1). Each will be discussed along with their symptoms.

Alcohol

Alcohol is known as a central-nervous-system depressant and it is the most frequently used drug in our society. Alcohol is the most familiar, and most abused depressant. With some exceptions, all depressants affect people in much the same way as does alcohol. Beer, wine, and various types of hard liquor make up this category.

Some obvious symptoms of alcohol abuse are slurred, mumbled, or incoherent speech; unsteady gait; relaxed social inhibitions; and lack of coordination. Other symptoms include impaired ability to divide attention; slowed reflexes; impaired judgment and concentration; impaired vision and coordination; and a wide variety of emotional effects such as euphoria, depression, suicidal tendencies, and laughing or crying for no reason. Sometimes the signs of alcohol abuse are very obvious. The smell of alcohol on a person's breath or clothing; a tired appearance, and clothing which seems to be out of place are signs associated with the condition called *hangover*.

Table 9–2 describes the effect of blood-alcohol concentration on physical and mental behaviors. As a protection officer, you may not encounter people driving vehicles or be in a position to test people through various types of drug alcohol methods, yet it is important to recognize these categories in terms of what the law defines as being under the influence of alcohol. As described in Table 9–2, the percentages range from 0.01 up to 0.30. As indicated, each particular level of alcohol concentration relates to a particular symptom. The 0.01 level, for example, is perhaps a small drink of alcohol which results in a slight tingling of the mucous membranes. Additional consumption would increase the symptoms.

Currently, the laws in most states clearly define the levels of 0.08 or 0.10 as the points of legal intoxication. What these percentages actually mean is that 0.10 translates basically to 1/10 of 1 percent of alcohol in your blood system. Of course, a 0.30 blood alcohol level is approaching 1/2 of 1 percent (0.50 would be 1/2 of 1 percent); people in this category often die due to an alcohol overdose. This means that in most states a person can be legally drunk, regardless of what the person may think! Therefore, at those levels a person is subject to arrest and prosecution. As an individual goes above 0.10, 0.20, to 0.30 levels in the table, the symptoms become much more severe, resulting in blackouts, stupor, and death.

Non-Prescription Stimulants

The major drugs abused in this category include cocaine and amphetamines. Street names used for cocaine are *coke, rock, crack, croak* and *spacebase*.

Table of Substance Abuse

Drug Used	Physical	Look For	Dangers
Alcohol (beer, wine, liquor)	Intoxication, slurred speech, unsteady walk, relaxation, relaxed inhibitions, impaired coordination, slowed reflexes.	Smell of alcohol on clothes or breath, intoxicated behavior, hangover, glazed eyes.	Addiction, accidents as a result of impaired ability and judgment, overdose when mixed with other depressants, heart and liver damage.
Cocaine (coke, rock, crack, base)	Brief intense euphoria, elevated blood pressure and heart rate, restlessness, excitement, feeling of well-being followed by depression.	Glass vials, glass pipe, white crystalline powder, razor blades, syringes, needle marks.	Addiction, heart attack, seizures, lung damage, severe depression, (see Stimulants).
Marijuana (pot, dope, grass, weed, herb, hash, joint)	Altered perceptions, red eyes, dry mouth, reduced concentration and coordination, euphoria, laughing, hunger.	Rolling papers, pipes, dried plant material, odor of burnt-hemp rope, roach clips.	Panic reaction, impaired short-term memory, addiction.
Hallucinogens (acid, LSD, PCP, MDMA, Ecstasy, psilocybin mushrooms, peyote)	Altered mood and perceptions, focus on detail, anxiety, panic, nausea, synaesthesia (ex: smell colors, see sounds)	Capsules, tablets, "microdots," blotter squares.	Unpredictable behavior, emotional instability, violent behavior (with PCP).
Inhalants (gas, aerosols, glue, nitrites, Rush, White out)	Nausea, dizziness, headaches, lack of coordination and control.	Odor of substance on clothing and breath, intoxication, drowsiness, poor muscular control.	Unconsciousness, suffocation, nausea and vomiting, damage to brain and central nervous system, sudden death.
Narcotics Heroin (junk, dope, Black tar, China white) Demerol, Dilaudid (D's) Morphine, Codeine	Euphoria, drowsiness, insensitivity to pain, nausea, vomiting, watery eyes, runny nose (see Depressants).	Needle marks on arms, needles, syringes, spoons, pin-point pupils, cold moist skin.	Addiction, lethargy, weight loss, contamination from unsterile needles (hepatitis, AIDS), accidental overdose.
Stimulants (speed, uppers, crank, Bam, black beauties, crystal, dexies, caffeine, nicotine, cocaine, amphetamines).	Alertness, talkativeness, Wakefulness, increased blood pressure, loss of appetite, mood elevation.	Pills and capsules, loss of sleep and appetite, irritability or anxiety, weight loss, hyperactivity.	Fatigue leading to exhaustion, addiction, paranoia, depression, confusion, possibly hallucinations.
Depressants Barbiturates, Sedatives, Tranquilizers, (downers, tranks, ludes, reds, Valium, yellow jackets, alcohol)	Depressed breathing and heartbeat, intoxication, drowsiness, uncoordinated movements.	Capsules and pills, confused behavior, longer periods of sleep, slurred speech.	Possible overdose, especially in combination with alcohol; muscle rigidity; addiction, withdrawal and overdose require medical treatment.

Source: 1991 "Drug Education Guide," The Positive Line #79930. Positive Promotions, 222 Ashland Place, Brooklyn, N.Y. 11217.

TABLE 9-2
Blood Alcohol Table

Blood Alcohol Concentration (%)	Physical and Mental Behavior
0.01	Slight tingling of mucous membranes.
0.02	Mild throbbing at the back of the head. A touch of dizziness. Personal appearance of no concern. Willing to talk.
0.03	Feeling of euphoria and superiority. ("Sure am glad I came to your party." "We will always be friends.")
0.04	Talking and laughing loudly. Movements a bit clumsy. Flippant remarks. ("You don't think I'm drunk, do you?")
0.05	Normal inhibitions almost eliminated. Many liberties taken. Talkative. Some loss of motor coordination.
0.07	Feeling of remoteness. Rapid pulse. Gross clumsiness.
0.10	Legally drunk in most states. Staggering, loud singing. Drowsiness. Rapid breathing.
0.20	Blackout level. Inability to recall events later. Easily angered. Shouting, groaning, weeping.
0.30 and above	Stupor. Breathing reflex threatened. Deep anesthesia. Death is due to paralysis of the respiratory center and is generally preceded by 5–10 hours of stupor and coma.

Some of the symptoms of cocaine abuse would be elevated blood pressure, respiration, and heart rate; restlessness; excitement; nervousness; irritability; inability to concentrate; and feelings of well-being followed by bouts of depression.

Cocaine is used in a variety of ways: smoking (freebasing), sniffing or snorting, injection, and oral consumption. Some of the instruments used with cocaine are glass vials, razor blades, crystalline powder, and syringes. Needle marks on a person's arms could indicate cocaine use.

Cannabis

Another category of drug is marijuana, a form of the cannabis plant. It is one of the more commonly used drugs in the work place. It is usually smoked. Common street names for this drug are *pot*, *dope*, *grass*, *weed*, *herb*, and *hash*. Symptoms of marijuana are altered perceptions, red eyes, dry mouth, and reduced concentration and coordination. Also, a burnt-hemp odor permeates the air when people are smoking marijuana.

Hallucinogens

Hallucinogens cause hallucinations (sensory experiences of things that do not exist outside the mind). This category of drugs include LSD, pilocybin and peyote. Their common names are *LSD, PCP*, and *acid*. These drugs are very unpredictable. They can cause mood swings, altered perceptions; people on these drugs display anxiety and panic. Some people under the influence of PCP engage in very violent behavior. Therefore, a person under the influence of a hallucinogen should be regarded as potentially dangerous.

Inhalants

Inhalants include a wide variety of breathable chemicals. This category of drugs includes gas, aerosols, glue, and nitrates. Widely abused volatile solvents include plastic cement or model airplane glue, paint, gasoline, lacquer thinners, and fingernail polish removers. Abused aerosols include hair sprays, deodorants, insecticides, glass chillers, and vegetable frying-pan lubricants. Less frequently used, but also in this category, are the anesthetic gases. These gases include ether, chloroform, amyl nitrite, butyl nitrite and nitrous oxide. Nausea, dizziness, headaches, and lack of coordination are symptoms of these types of drugs. Persons under the influence of inhalants will generally appear confused and disoriented and their speech will usually be slurred.

Narcotics

This category includes, heroin, junk, dope, codeine, and Demerol. These drugs create drowsiness, sensitivity to pain, vomiting, and watery eyes. They are oftentimes referred to as *pain-killing drugs*; morphine is one category of narcotics. Some of these narcotics can be legally prescribed. Morphine, for example, was first used as a pain killer in 1825. The invention of the hypodermic needle syringe in 1843 increased its use and eventual abuse.

Prescription Stimulants

Some of the street names for stimulants include *speed, uppers, crank, crystal, meth, black beauty. dexies*, and *bennies*. Alertness, talkativeness, wakefulness, increased blood pressure, and loss of appetite are all examples of symptoms exhibited with the abuse of this type of drug. These are usually produced in pill, capsule, tablet, and liquid-elixir form. They are taken orally.

Depressants

Depressants are drugs which lower the rate of muscular or nervous activity. They have the effect of increasing drowsiness and uncoordinated movements.

This category includes barbiturates, sedatives, antidepressants, and antipsychotic tranquilizers. Some of the street names are *downers, tranks, ludes, reds*, and *valium*.

GUIDELINES FOR RESPONSE

Up to this point the discussion has focused on some of the types of drugs and the problems of drug abuse in society. The question raised is: What is the protection officer expected to do when encountering a person under the influence of a drug? These encounters may occur in the work place or in other public settings such as a parking lot.

The basic rule when responding to people under the influence of drugs is not to antagonize or anger the person. This means, keep away from the person and observe the individual to see what he or she is doing. Be very cautious when approaching a person who may be under the influence of drugs. Check to make sure that the person has no weapons and watch the persons' movement, especially the hands. Try to reason with the person (although this may be difficult). Ask if they need assistance or if they want you to notify someone. Remember, a drug abuser could also have a medical problem requiring immediate attention. Moreover, a person could be epileptic or diabetic and their symptoms may resemble someone under the influence of drugs. Check to see if the person is wearing identification indicating a diabetic or epileptic condition.

Review your company policy or check with the supervisors to make sure that you understand what you are supposed to do when encountering someone who may be under the influence of drugs. The following are offered as general guidelines:

1. If you observe an employee who may be under the influence, or using a drug in the work place, contact your supervisor immediately! Document

the behavior of this individual by writing down anything you have observed the person doing, what they may have in their possession, as well as the kind of behavior they are displaying (how the person was dressed, general appearance, coordination, and manner of speech).

2. If you encounter a person under the influence of drugs, whether an employee or invitee, and the person is not being cooperative (for example, the person will not leave), call for assistance and do not attempt to physically control the person alone! Also, attempt to get a witness to observe this person and who can also act as a safety measure in case the person becomes violent. Use verbal skills to control the individual. Make every effort to keep the person away from other people.

3. Remove the person to an office or to a place out of view of the public. Make sure the person does not have instruments or anything that may be used as a potential weapon. Check the person for weapons or other instruments likely to cause harm. Your safety and the safety of others is important in these situations.

4. If you encounter someone who is unconscious or very ill, call for medical assistance immediately! Deaths can result from drug overdoses; therefore, it is very important to contact medical assistance as well as the supervisor when suspecting that a person may be very ill or unconscious. If the person complains of injury or of being very ill, do not move him or her. Make the person as comfortable as possible and call for medical assistance immediately. Check for necklaces or wrist bands that may suggest a medical problem (diabetes, epilepsy).

5. If you find narcotics or narcotic instruments such as syringes, try not to pick them up. Do not move them from the place where you found them. However, if the items must be moved, pick them up gently using some type of cloth or protective device so you will not be harmed or the evidence destroyed. Put the items in a plastic bag and take them to a secure area for further investigation. The items uncovered may be used in court at a later date.

Conclusion

It is becoming increasingly common to encounter employees or others who may be abusing drugs. The use of illegal drugs is not only a social problem but a business problem. Drug abuse is not limited to young people; persons of all ages and occupations are abusing drugs. Police around the country have already started to see its impact in terms of crime increases. In the field of private protection, the identification of substance abusers is of the utmost importance. Be aware of suspicious persons who may be loitering in the area. These persons may be drug dealers. If such persons are observed, attempt to get a description and notify the proper authorities. Because the use of drugs in the work place is becoming a major problem, attempts are being made to identify users through preemployment drug screening and undercover investigations.

Glossary of Drug Slang Terms

A

Acid Head	User of LSD
Airhead	Under the influence of marijuana

B

Bad Trip	Unpleasant episode with an hallucinogen
Bag Man	Person who transports money
Bang	To inject narcotics
Big Man	Supplier of drugs

Bindle	A small packet of drug powder
Blanks	Low-quality drugs
Blasted	Under the influence of drugs
Blow	Snort or sniff cocaine or smoke marijuana
Bong	A cylindrical water pipe used to smoke marijuana
Bread	Cash
Broker	Go-between for a drug deal
Bummer	A bad experience with drugs
Burned	Purchase nongenuine drugs
Burnout	Heavy abuser of drugs
Busted	Arrested
Buzz	Under the influence of drugs

C

Caps	Drug capsules
Charged Up	Under the influence of drugs
Chippy	Person who uses drugs infrequently
Chipping	Occasional use of drugs
Coasting	Under the influence of drugs
Cold Turkey	Sudden withdrawal from drugs
Connect	To purchase drugs
Connections	Supplier of drugs
Cop	To obtain drugs
Crash	Sleep off effects of drugs
Cut	To adulterate drugs
Cut Out	To leave from some place

D

Dealer	A seller of illegal drugs
Deck	A packet of drugs
Dime Bag	Ten-dollar bag of drugs
Dope	Drugs
Drop	To take drugs orally, or a place where money or drugs are left
Dynamite	High-quality or potent drugs

E

Easy Score	Obtaining drugs without difficulty
Eighth	One-eighth of a pound of drugs

F

Factory	Place where drugs are diluted, packaged or manufactured
Fall	Arrested
Feds	Federal agents
Fix	Inject drugs
Flashback	Reoccurrence of hallucinations
Flea Powder	Poor-quality drugs
Flying	Under the influence of drugs
Freebasing	Smoke cocaine through a special water pipe

Freeze	To renege on a drug transaction
Front	To put money out before receiving the merchandise
Fuzz	Police

G

Goods	Drugs
Gram	A metric measure of weight
Gun	Equipment for injecting drugs

H

Hand-To-Hand	Direct delivery and payment
Heat	Police (or gun)
Head Shop	Store that specializes in the sale of drug paraphernalia
Heeled	Having plenty of money
High	Under the influence of drugs
Hit	A single does of drugs
Holding	In possession of drugs
Hooked	Addicted
Hopped Up	Under the influence of drugs
Hot	Wanted by authorities
Hot Shot	Injecting an overdose of drugs
Hustle	Attempt to obtain drug customers
Hype	Heroin addict

I

Ice Cream Habit	Occasional use of drugs
In	Connected with drug suppliers

J

Jag	Under the influence of drugs or alcohol
Jive	Drugs
Joy Popping	Occasional use of drugs
Junkie	Narcotic addict

K

Key	Kilogram
Kick	Getting off a drug habit
Kiddie Dope	Usually prescription of drugs
Kit	Equipment to inject drugs

L

Lemonade	Poor-quality drugs
Lettuce	Money
Lid	One ounce or less of marijuana
Line	Cocaine arranged in a row
Load	A large quantity of drugs

M

Mainliner	A person who injects directly into the veins
Meet	Buyer and seller get together
Man	Police
Merchandise	Drugs

Monkey	Drug dependency
Mule	A carrier of drugs

N

Nailed	Arrested
Narc	Narcotic agent
Needle Freak	A person who prefers to take drugs with a needle
Nickel Bag	Five-dollar bag of drugs

O

OD	Overdosed on drugs
On A Trip	Under the influence of drugs
On Ice	In jail
On The Bricks	Walking the streets
On The Nod	Under the influence of narcotics or depressants
Out Of It	Under the influence of drugs
O.Z.	One ounce

P

Panic	Drugs not available
Pepsi Habit	Occasional use of drugs
Pickup	Purchase drugs
Piece	Usually one ounce of drugs
Plant	A hiding place for drugs
Pot Head	Marijuana user

Q

Quack	Doctor

R

Rap	Criminally charged, or to talk with someone
Riding The Wave	Under the influence of drugs
Rig	Equipment for injecting drugs
Roach Clip	A device used to hold the butt of a marijuana cigarette
Rollers	A term used by a lookout and yelled when police come
Rush	A sudden intense, euphoric effect from taking drugs

S

Scene	A special location or condition
Score	Purchase drugs
Scratch	Money
Script	A doctor's prescription
Shoot Up	To inject drugs
Shooting Gallery	Place where drugs are used
Skin Popping	Shooting (injecting) drugs under the skin
Slammer	Jail
Snort	To sniff drugs
Space Cadet	Under the influence of drugs
Speed Freak	Habitual user of methamphetamine
Spike	Needle (syringe)
Split	To leave from someplace

Stash	Place where drugs are hidden
Step On	To dilute drugs
Stoned	Under the influence of drugs
Straight	Not using drugs
Strung Out	Heavily addicted to drugs

T

Take A Powder	To leave or get lost
Taste	A small sample of drugs
Toke	Inhaling marijuana or hashish smoke
Toot	To sniff cocaine
Tracks	A row of needle marks on a person
Trap	Hiding place for drugs
Trip	Under the influence of drugs
Turf	A location where drugs are sold
Turkey	Nongenuine drugs
Turned On	Introduced to drugs, or under the influence of drugs

U–V

Uncle	Federal agents

W

Wasted	Under the influence of drugs; also murdered
Works	Equipment for injecting drugs

Z

Zombie	Heavy user of drugs

Discussion Questions

1. What are the five warning signs of drug-dependent persons?
2. List the eight categories of drugs and give an example of each.
3. What is the most commonly abused drug in the work place?
4. If you encounter someone who is unconscious but breathing, and you detect the odor of alcohol, what should you do?
5. What medical conditions have some of the same symptoms as a person under the influence of a drug?

References

Digest: Drugs of Abuse, Institute for Substance Abuse Research, Vero Beach, Florida, 1985.

DRUG EDUCATION GUIDE, The Positive Line #79930, Brooklyn, NY: Positive Promotions, 1991.

HESS, FAREN, AND W. BENNETT, *Criminal Investigation* (3rd ed.) St. Paul, MN: West Publishing, 1991.

KELLY, JACK, "Poll: On-Job Drug Use Is Significant," *U.S.A. Today*, December 16, 1989.

LISKA, KEN, *Drugs and the Human Body*. New York, NY: MacMillan Publishing Co., 1990.

TIMM, HOWARD W., AND KENNETH E. CHRISTIAN, *Introduction to Private Security*. Pacific Grove, CA: Brooks/Cole Publishing Company, 1991.

U. S. DEPARTMENT OF JUSTICE, *N.J. Reports*. Washington, D.C.: Government Printing Office, 1989.

Code of Ethics for Private Security Employees

In recognition of the significant contribution of private security to crime prevention and reduction, as a private security employee, I pledge

1. To accept the responsibilities and fulfill the obligations of my role: protecting life and property; preventing and reducing crimes against my employer's business, or other organizations and institutions to which I am assigned; upholding the law; and respecting the constitutional rights of all persons.

2. To conduct myself with honesty and integrity and to adhere to the highest moral principles in the performance of my security duties.

3. To be faithful, diligent, and dependable in discharging my duties, and to uphold at all times the laws, policies, and procedures that protect the rights of others.

4. To observe the precepts of truth, accuracy, and prudence, without allowing personal feelings, prejudices, animosities, or friendships to influence my judgments.

5. To report to my superiors, without hesitation, any violation of the law or of my employer's or client's regulations.

6. To respect and protect the confidential and privileged information of my employer or client beyond the term of my employment, except where their interests are contrary to law or this code of ethics.

7. To cooperate with all recognized and responsible law enforcement and government agencies in matters within their jurisdiction.

8. To accept no compensation, commission, gratuity, or other advantage without the knowledge and consent of my employer.

9. To conduct myself professionally at all times, and to perform my duties in a manner that reflects credit upon myself, my employer, and private security.

10. To strive continually to improve my performance by seeking training and educational opportunities that will better prepare me for my private security duties.

B

Hourly Wages for Protection Officers for Selected Cities

Metropolitan Area	Mean	Median	Middle Range
Anaheim, CA			
Manufacturing	$10.37	$11.05	$6.88–13.56
Nonmanufacturing	5.51	5.48	4.80–5.75
Atlanta			
Manufacturing	11.29	13.85	7.75–14.68
Nonmanufacturing	5.61	5.25	5.00–5.75
Austin			
Manufacturing	7.31	7.54	6.76–7.89
Nonmanufacturing	4.50	4.25	4.00–5.00
Baltimore			
Manufacturing	11.61	12.10	9.76–13.73
Nonmanufacturing	5.17	5.40	4.00–6.00
Boston			
Manufacturing	9.89	10.00	8.39–11.32
Nonmanufacturing	6.38	6.00	5.55–6.75
Charleston, SC			
Manufacturing	7.85	7.53	6.81–8.63
Nonmanufacturing	4.23	3.88	3.63–5.00
Charlotte, NC			
Manufacturing	6.43	5.84	5.36–6.33
Nonmanufacturing	5.69	5.00	4.20–6.00
Chicago			
Manufacturing	9.37	8.86	6.00–12.87
Nonmanufacturing	6.01	5.25	4.80–6.85
Cincinnati			
Manufacturing	10.53	10.80	8.85–12.84
Nonmanufacturing	4.20	4.00	3.75–4.25
Cleveland			
Manufacturing	9.97	10.56	7.72–12.32
Nonmanufacturing	4.60	4.25	4.00–4.75

Metropolitan Area	Mean	Median	Middle Range
Dallas			
Manufacturing	9.55	8.75	7.52–11.75
Nonmanufacturing	5.15	4.75	4.50–5.50
Denver			
Manufacturing	10.67	13.30	7.05–13.38
Nonmanufacturing	4.86	4.40	4.00–5.00
Detroit			
Manufacturing	12.48	14.39	10.86–14.39
Nonmanufacturing	6.31	4.56	4.25–6.11
Hartford, CT			
Manufacturing	9.33	9.45	8.43–10.29
Nonmanufacturing	6.47	6.00	5.50–7.00
Houston			
Manufacturing	9.10	8.30	7.08–10.73
Nonmanufacturing	5.50	5.10	4.70–6.00
Huntsville, AL			
Manufacturing	5.65	5.62	4.80–6.16
Nonmanufacturing	4.19	4.00	3.50–4.30
Indianapolis			
Manufacturing	11.33	12.07	8.69–14.10
Nonmanufacturing	4.86	4.00	3.75–5.12
Kansas City			
Manufacturing	11.28	11.49	10.25–11.96
Nonmanufacturing	4.56	4.00	3.65–4.75
Los Angeles			
Manufacturing	10.54	10.35	8.50–13.56
Nonmanufacturing	5.61	5.15	4.75–6.10
Miami			
Manufacturing	6.92	7.02	6.37–7.64
Nonmanufacturing	5.21	5.00	4.35–5.72
Milwaukee			
Manufacturing	13.41	14.34	14.34–14.34
Nonmanufacturing	4.67	4.25	3.90–5.09
Minneapolis			
Manufacturing	10.90	11.06	10.28–11.18
Nonmanufacturing	5.47	5.25	4.75–5.75
Newark			
Manufacturing	11.78	11.18	10.75–13.87
Nonmanufacturing	5.49	5.00	4.50–6.16
New Orleans			
Manufacturing	8.00	8.22	6.76–9.31
Nonmanufacturing	4.36	3.73	3.50–4.20
New York			
Manufacturing	11.00	11.62	9.14–13.07
Nonmanufacturing	5.80	5.00	4.56–6.10
Philadelphia			
Manufacturing	10.99	10.73	9.79–12.45
Nonmanufacturing	5.18	5.00	4.33–5.50

continued on next page

Metropolitan Area	Mean	Median	Middle Range
Phoenix			
Manufacturing	8.16	8.35	6.15–9.83
Nonmanufacturing	4.99	4.65	4.50–5.00
Pittsburgh			
Manufacturing	10.60	10.62	8.93–11.77
Nonmanufacturing	4.73	4.27	3.75–5.25
Riverside, CA			
Manufacturing	7.75	7.47	5.42–9.63
Nonmanufacturing	4.96	4.70	4.25–5.40
Salt Lake City			
Manufacturing	6.61	6.50	5.95–7.20
Nonmanufacturing	4.43	4.50	3.75–5.00
San Jose, CA			
Manufacturing	10.54	10.23	8.42–13.99
Nonmanufacturing	6.43	6.25	5.80–7.00
St. Louis			
Manufacturing	12.12	12.76	10.31–13.85
Nonmanufacturing	4.45	4.00	3.80–4.75
Washington, D.C.			
Manufacturing	8.75	7.89	7.12–10.25
Nonmanufacturing	7.02	7.25	6.13–7.45

Notes: Mean or average is computed by totaling earnings of all the workers and dividing by the total number of workers. The median designates the wage at which half of the workers earn more and half earn less than the amount stated. The middle range is derived from two values: a fourth of the workers receive the same or less than the lower of these two rates and a fourth earn the same or more than the higher rate.

Source: Fischer, Robert J., and Gion Green. *Introduction to Security* (5th ed.) Boston, MA: Butterworth-Heinemann, 1992, pp.83–85.

Appendix

C

180-Hour Five-Week Security Officer Course*

INTRODUCTION   2 Hours

SECURITY OFFICER
"POWERS TO ARREST"   8 Hours
Historical Overview of Private
    Security Industry
Types of security; process of becoming
    a registered officer
Role and responsibilities of today's
    security officer
Criminal and civil liabilities
Arrest procedures; limitations
Interaction/cooperation with public
    law enforcement & other agencies
State examination for certification; fin-
    gerprinting, etc.

PROFESSIONALISM
& ETHICS   8 Hours
Security awareness, education, and
    training
Legislation, court action, and other
    elements affecting law enforcement
    & security
Community, public, and media relations
Changing role of today's security pro-
    fessional
Importance of honesty, respect, punc-
    tuality, and accountability

Industry trends...potential future
    events

CRIMINAL LAW   8 Hours
Historical overview and theory
    of law
The criminal justice system
Courtroom procedures
Crime definitions and elements

RETAIL SECURITY   4 Hours
Shopping services
Shoplifting (Merchant's Retail Theft
    Law - 490.5 Penal Code)
Internal theft(s)
Asset protection
Checks, credit cards, and other
    "paper crimes"

BATON   8 Hours
Authority and regulatory considera-
    tions for baton use
Use-of-force and liability issues
Fundamentals of baton handling
First aid for baton injuries

TEAR GAS COURSE   2 Hours
Reading supplement for tear gas
    course
Video presentation

*Public Safety Training Association, Inc., San Diego, CA

CPR (CARDIOPULMONARY RESUSCITATION) & FIRST AID CONCEPTS — 8 Hours

Basic support for witnessed and unwitnessed cardiac arrests

External cardiac compression techniques

Special resuscitation situations

Artificial respiration and artificial circulation

Fundamental first aid methods and techniques

INDUSTRIAL SECURITY — 4 Hours

How security operates in a hotel, hospital, & industrial-type setting

Specific problems and solutions

Overview of asset protection

Interior and exterior controls

Video: "High-rise Fire Safety"

SECURITY & SAFETY PATROL PROCEDURES — 11 Hours

Duty preparedness and assumption

Observation, perception, surveillance techniques

Vehicle safety and operations

Field contacts (criminal and noncriminal)

Criminal psychology

Traffic and crowd control

Radio communications

Gangs, groups, and cults

Disaster preparedness (Earthquake Handouts) Video - "Surviving the Big One"

Peace Officers/Security Officers Face the AIDS Crisis (Video)

FIRE SAFETY, HAZARDOUS MATERIALS AND BOMB PLAN PROCEDURES — 4 Hours

Definition of fire, classes of fire, and methods to fight fires

Basic aids to help the security officer in determining origins and approximate durations of fires involving structures

Types of hazardous materials and today's work place

Basic bomb threat procedures

DISCRETIONARY DECISION MAKING — 3 Hours

Understanding of importance of self-esteem, exercising good communication skills, and wise use of technological equipment

Problem definition, possible solutions and alternatives

Unrelated barriers, timely use of discretion

"Stressors" that influence decision-making process

SECURITY AND ALARM SYSTEMS — 3 Hours

Types of security alarm systems

Advantages and disadvantages of alarm systems

Demonstration of portable alarm systems

WORK PREPAREDNESS TRAINING — 12 Hours

Psychological testing

Reading comprehension testing

Expectations, attitudes, self-esteem, literacy, and self motivation

Personal recordkeeping and resume development

Background checks and employer "wish" list

Job applications, interview techniques and employment referrals

COMMUNICATION SKILLS — 11 Hours

Walkie-talkie, beeper, word processor and FAX - expectations and limitations

Verbal and nonverbal communications

Field note taking

Mechanics of report writing

Semantics and other barriers to effective report writing

FIREARMS — 16 Hours

Moral and legal aspects of firearms usage

To arm or not to arm for security

Liability issues; alternatives to deadly force

Firearms nomenclature, sight and target pictures

Range rules, training, and certification procedures

Cleaning methods; first aid procedures

RESPONSES TO ASSAULTIVE BEHAVIORS — 5 Hours

Alternatives to deadly force situations

Weaponless self-defense

Handcuffing procedures

SHOPPING CENTER SECURITY — 3 Hours

Overview of shopping center growth and related problems

Security manual and emergency planning

Liaison with retail security and public law enforcement

DRUGS IN THE WORKPLACE — 4 Hours

The problem of drugs in society today

Impact of substance abuse in the work place

Common drugs of abuse

Legal considerations for employers and security personnel

PRINCIPLES OF ORGANIZATION AND SUPERVISION — 6 Hours

The paramilitary structure of law enforcement and security

Unity of command, span of control, manpower distribution

Rules, regulations, and policy manuals; following directions

Contracts; mathematically computing cost estimates, and manpower needs

Labor laws

INTRODUCTION TO PRIVATE SECURITY — 32 Hours

Private Investigator's Act-A Beginning

Investigative techniques

Public records; sources of information

Writing for effectiveness

Personal protection

APARTMENT-CONDO SECURITY — 3 Hours

Apartment-Condo living; reality in the 90s

Landlord-tenant disputes

Unique problems and possible solutions

RAILWAY & TROLLEY SECURITY — 3 Hours

Role and responsibilities of the railway-trolley inspector and security officer

The laws creating light rail systems and enforcement of those laws

Problems and solutions

AIRPORT SECURITY — 3 Hours

Airport security and safety problems

Inspections, searches and seizures; legal considerations

Parking problems

Threat of terrorism

CULTURAL, ETHNIC, & DOMESTIC RELATIONS — 5 Hours

Domestic violence—criminal and/or civil?

Common sex crimes

The security officer's role when confronted with domestic violence and sex-related offenses

INTRODUCTION TO CORRECTIONS — 3 Hours

Corrections and prisons as a career opportunity

Problems and solutions

STAFF TIME — 3 Hours

Introduction of course by PSTA staff

Rules, regulations and policies of school & regulations

Appendix
D

State Statutes Regulating
Security Guards*

State	Registration	In-House Exclusion	Requirements	Liability Requirement	Remarks
Alabama	None				Local licensing
Alaska	Yes			Bond	$25.00 per guard
Arizona	Yes			$300,000	Local licensing
Arkansas	Yes	Unarmed only	Exam	$100,000	Exam administered by trainer. Exam & 2 yrs. experience required for trainer.
California	Yes	Yes	Exam	Bond	Powers to arrest
Colorado	Yes			Local licensing	
Connecticut	Yes			Bond	
Delaware	Yes			$10,000 bond	
Florida	Yes			$100,000 per person, $300,000 per occurrence	
Georgia	Yes	Pending	8 hrs. classroom instruction	$25,000 bond	Can work 30 days before training
Hawaii	Yes			$5,000 bond	
Idaho	None				
Illinois	Yes				New reg. pending. House Bill 643
Indiana	Yes	Yes			
Iowa	Yes	Yes	Local exam administered by law enforcement	$2,000 bond	
Kansas					Local requirements
Kentucky	None		Local	None	House Bill 367—state requirements pending
Louisiana			Local		

*Compiled by Minot B. Dodson, CPP.

State	Registration	In-House Exclusion	Requirements	Liability Requirement	Remarks
Maine	Company			Bond	
Maryland	State ID card			$5,000 bond	Company must be licensed as PI agency
Massachusetts				$5,000 bond	Company must maintain records subject to audit
Michigan				$10,000 bond, L$25,000/$100,000/$200,000	
Minnesota		Yes		Bond	
Mississippi			Local		
Missouri	Local		Local 3 days training & exam		
Montana	Yes				New legislation pending
Nebraska	Local				
Nevada	Yes			$325,000L	
New Hampshire	Yes	Yes		Bond	
New Jersey	Yes			$5,000 bond	
New Mexico	Yes	Yes		Bond	
New York	Company				
North Carolina	State ID card			$50,000/$100,000/$200,000	
North Dakota	Yes	Yes	Training	Bond	New legislation took effect 1/1/84
Ohio	Yes	Yes		$100/$300,000	
Oklahoma			Local		
Oregon			Local		
Pennsylvania			State police check	$10,000 bond	Company must be licensed as PI agency
Rhode Island	None				
South Carolina	Yes		4 hrs. training	$10,000 bond	
South Dakota	None				
Tennessee			Local		
Texas	Yes	Yes		Bond	New legislation took effect 1/1/84
Utah	Yes		Training	$300,000L	Training administered by state qualified agent
Vermont	Yes		Exam	Bond	
Virginia	Yes	Yes	Training (12 hrs)	$5M bond or L-$100/300M	120 days to complete training
Washington			Local		
West Virginia	Yes	Yes	Training	$2500 surety bond	Employer training requirements approved by state
Wisconsin	Yes			$10M-L	
Wyoming			Local		

Note: Since this table was produced, a number of states have modified or are considering changes in training and registration requirements for protection officers. Officers are urged to check with their jurisdictions to verify standards.

Index

173

Equipment, patrol, 35–36
Evidence, 43–45
　　circumstantial, 43–45
Excessive force, 25
Exclusionary rule, 23
Executive branch of government, 14

F

Facial expressions, as nonverbal
　　　　communications, 80
False arrest, 25
False imprisonment, 20, 25
FBI Bomb Data Center, 125, 126
Federal Aviation Administration (FAA),
　　117
Felony, defined, 15
Finger sweep, 140
Firearms, 110–11
Fire prevention, 120–24
　　extinguishers, 122–24
　　　　types of, 123–24
　　procedures for, 124
　　response procedures, 124
Fires, 132
First aid/CPR, 135–42
　　infectious disease precautions, 136
　　legal aspects of, 136
　　phoning for emergency medical
　　　　　services, 138–39
　　special emergencies, 139–42
　　　　bleeding, 141–42
　　　　burns, 145
　　　　choking, 139–40
　　　　dislocations, 145–46
　　　　finger sweep, 140
　　　　fractures, 145–46
　　　　poisoning, 143–45
　　　　rescue breathing, 139
　　　　shock, 142–43
　　　　sprains, 145–46
　　　　strains, 145–46
　　surveying the scene, 136–37
　　surveying the victim, 137–38
First-degree burns, 145
Floodlights, 99
Floods, 132
Foam, as fire extinguisher, 123
Foil, as detector, 102
Food and Drug Administration (FDA),
　　117
Food Safety and Quality Service, 117
Foot patrol, 36–37
Force, 105–15
　　encounters, patterns of, 111–13
　　excessive, 25
　　handcuffing, 113–14
　　levels of, 106–7

　　reasonable, justifying, 106
　　weapons, 107–11
　　when to use, 107
Forest fires, 131
Fractures, 145–46
Fresnal units, 99

G

Gaseous-discharge lamps, 100
Gestures, as nonverbal communications,
　　80
Glass break sensors:
　　piezoelectric, 103
　　vibration, 102
Globe Security, 5
Government defense contractor,
　　　　protection for, 9
Government scientific laboratory,
　　　　security duties in, 6
Grammar, 52

H

Hallucinogens, 155, 156
Halogenated agents (Halon), as fire extin-
　　　　guisher, 123–24
Handcuffing, 113–14
Hazardous materials, 127–31
　　hazardous chemicals, 128
　　placards, 129
　　procedures, 129–31
　　subclassifications, 127
Health and safety, 116–49
　　bomb threats, 124–27
　　disasters, 131–35
　　fire prevention/response, 120–24
　　first aid/CPR, 135–42
　　hazardous materials incidents, 127–31
　　laws, 117–20
Heimlich maneuver, 140
Hiring, negligent, 29
Hurricanes, 131

I

Incandescent lamps, 100
Incident report:
　　parts of, 70
　　writing, 67–72
　　　　guidelines for, 71–72
In-court testimony, 46–47
Indecent exposure, 17
Independent clauses, 54